D0961852

HEREDITY AND HOPE

HEREDITY AND HOPE

The Case for Genetic Screening

RUTH SCHWARTZ COWAN

Harvard University Press
Cambridge, Massachusetts
London, England
2008

Printed in the United States of America

Library of Congress Cataloging-in-Publication Data

Cowan, Ruth Schwartz, 1941–
Heredity and hope : the case for genetic screening /
Ruth Schwartz Cowan.
p. cm.
Includes bibliographical references and index.
ISBN-13: 978-0-674-02424-3 (alk. paper)
1. Genetic screening—History. 2. Eugenics—History.
I. Title.
[DNLM: 1. Genetic Screening—history. 2. Eugenics—
history. 3. History, 20th Century. QZ 11.1 C874h 2008]
RB155.65.C69 2008
616′.042—dc22 2007039201

For

Jennifer, May, Sarah—and all their children.

L'chaim!

Contents

Introduction *1*

1 Many Varieties of Beautiful Inheritance *12*

2 Eugenics and the Genealogical Fallacy *41*

3 Pronatal Motives and Prenatal Diagnosis *71*

4 No Matter What, This Has to Stop! *117*

5 Genetic Screening and Genocidal Claims *150*

6 Parents, Politicians, Physicians, and Priests *181*

Conclusion *223*

Notes *247*

Further Reading *269*

Acknowledgments *277*

Index *281*

HEREDITY AND HOPE

Introduction

In the late summer of 1979, I found myself lying on a gurney in a hospital, four months pregnant and very, very uncomfortable. I had just consumed eight liters of water, which, I was told, would make it easier for a doctor to "see" (with something called an ultrasound device) the baby that was growing in my uterus, so that some kind of test could be "performed" safely. I was thirty-eight; the nurse-midwife to whom I went for prenatal care had told me that, given my age, I was at increased risk of having a baby with Down syndrome and I ought to have "this new test" to find out whether I would.

Nothing she said (or, probably, could have said) prepared me for the weird and disturbing experience of genetic testing through amniocentesis: the initial discomfort; the dismay of being asked to read and sign a four-page consent form which laid out, in dull detail, all the possible hazards of false positives and false negatives; the terror of watching a man put a needle through my belly into the protected space in which my baby (only the doctor would have then called it a fetus) was developing; the shock of being told that that blur on the television monitor "was" that baby; and—worst of all—the interminable wait for results, the weeks of troubled sleep and anxious conversations with my husband while we tried to decide what we would do if the results came back "positive."

My research for this book began right after I signed that four-page consent form. I recall insisting that, despite my discomfort and despite the doctor's busy schedule, I was going to "read the whole damn thing" before signing. I was a feminist with a Ph.D.; this was 1979; *informed* consent was virtually my middle name, especially with regard to gynecological and obstetrical matters. The form was full of tables and statistically loaded phrases like "increased risk" and "low probability." I could understand (most of) it; I had, after all, written my dissertation about one of the founders of biostatistics. But what did women who did not have my background manage to make of it, women who could not tell a Gaussian distribution from a Fourier expansion? How could anyone who did not know the ins and outs of statistics make any sense of this form? As a historian of science and technology, I thought, I ought to try investigating the forces that had conspired to land me—and many other women like me—on that gurney, faced with that form, about to experience that awful (and awe-ful) genetic test.

Genetic testing is much more pervasive today than it was thirty years ago; it is also more complex and, if anything, more troubling. I have observed those changes within my own immediate family. Unbeknownst to me, my first two children, born in 1970 and 1973, were subjected to genetic tests as newborn babies. The state in which we lived (New York) required every hospital to take a heel-prick sample of blood from newborns (without asking for parental consent) and to have that sample tested (without charge to the parents) for the presence of a chemical that signaled the presence of a genetic disease, phenylketonuria (PKU). A generation later my grandchildren were similarly tested (still without parental consent, still with the heel-prick blood sample, still without direct charge to their parents), not just for PKU but for two dozen additional genetic diseases, some of which had been totally unknown to medical science when their parents were born.

At the time that my third child was assessed *in utero*, medical geneticists could identify only a few congenital conditions in a fetus (Down syndrome, some of the sex chromosome anomalies, and some of the neural tube defects, such as spina bifida), and only one kind of test (amniocentesis) was offered to me. By the time my oldest daughter was pregnant, twenty years later, she and her husband were offered a choice among five very different diagnostic procedures, which test for a much larger number of diseases and disabilities. At each step of this testing regimen, they were faced with a set of mind-boggling decision trees in which the likelihood of miscarriage or other damage to the fetus had to be balanced against the incidence of the disease in their ethnic groups, the chances that the fetus would be afflicted, the severity of the disability, the likely accuracy of the diagnostic test, and their willingness to consider the termination of the pregnancy for the particular condition in question. Three years later another daughter and son-in-law faced even more complicated decisions because she was carrying twins, a situation that doubles the likelihood of a problematic diagnosis at the same time as it lessens the predictive accuracy of many of the tests.

Today genetic decisions are not restricted to the results of fetal or newborn testing. Not long ago my mother died after a lengthy struggle with Alzheimer's disease. As a result of research done for the Human Genome Project, a mutation that causes some cases of Alzheimer's has been identified. My mother was never tested for that mutation; she was already well into her decline when it was discovered. I have no interest in being tested, partly because the test is unreliable and partly because there is neither a cure nor even a particularly effective therapy for Alzheimer's. But other adult-onset disease mutations have recently been discovered. If my mother had died, earlier in her life, from one of those diseases (say, breast or colon cancer), I might well decide to have myself tested, and I might advise my daughters to do likewise. And if they tested positive for the implicated gene, they would again confront many difficult decisions, first about preemptive surgery and then, possibly, also about prenatal testing. Parents who receive positive test results in such circumstances have to ask themselves: Is it right to terminate a pregnancy because the fetus carries a mutation that will increase the risk of (not necessarily cause, just increase the risk of) a painful and maybe fatal disease that will not, however, wreak its havoc until the person that fetus will become has lived for twenty or thirty reasonably healthy years?

When I was young there was an expression that my friends and I often used when we hit a personal roadblock or faced difficult times or troubling decisions. "No one," we used to say, "ever promised us a rose garden." True enough, but our families, our history, our cultural canon had at least given us some role models on which to try to pattern our responses when confronted by thorny situations. Death of a loved one, poverty, illness, divorce, disappointment, frustration—these had all been dealt with, and sometimes overcome, by someone we knew or someone we had read about.

Genetic testing promises us a rose garden—the prevention of devastating diseases and profoundly disabling conditions—but, unfortunately, there are precious few role models to help us make the unprecedented decisions that the testing forces upon us. Our parents and grandparents cannot be our models because these diagnostic tests did not exist when they were having children. The medical experts who give us the results are trained to refuse our requests for advice; they must lay out the options and then leave us, in the name of patient autonomy, to make the decisions. We hesitate to ask our friends for fear that they will think less of us and stop being our friends. Our religious texts say not a word about what it means to be a good person or to serve God's will in these extraordinary situations; our religious advisors, if we speak to them at all, seem to be mouthing platitudes. Few novels we have ever read, or films we have viewed, or television programs we have watched, have been concerned with how best to make these gut-wrenching decisions.

And gut-wrenching is precisely what they are. One minute we are filled with joy as we anticipate the advent of a new life; the next minute the tests results are in and we are filled with terror as we anticipate deciding to end that new life before it has properly begun. On top of this our decisions must be made quickly—in a few days or, if we are lucky, a few weeks. In that time we must think deeply about our most intimate, emotion-laden relationships, the ones we usually push to the back of our consciousness just to be able to get through the day: relationships between sexual partners, between spouses, between parents and children. The decisions we make about genetic testing depend on our answers to questions on which even philosophers and theologians disagree. What is a good life? What is a good parent? What is good health—and how important is it? What does it really mean to love a fetus, a child, a spouse, yourself?

Of course these are not just personal questions; they are social and political questions as well. Every reproductive decision, every reproductive event, is as much social as it is personal. Every baby creates obligations and opportunities for the society that it joins, and, reciprocally, every society confers rights as well as responsibilities on every parent of every baby. The particulars may differ (in some societies, parents have the right to choose spouses for their children, in other societies they don't; some societies require parents to send their children to school, others do not), but the fact that parenthood is a social role—and therefore also economic and political—is universally true. That is why some babies have social security numbers before they even have names.

Since the early 1960s medical geneticists have been trying to improve children's health and parents' lives by interfering, in one way or another, with reproductive decisions. Many people have come to believe that prevention of disease through genetic testing is a very worthy enterprise; in most developed countries of the world there has been a great deal of social, political, and economic support for this project.

Millions upon millions of dollars—some private, most public—have been allocated to the research that has led to the tests. Legislators in hundreds of jurisdictions have mandated that all newborn babies be tested, at public expense, for a variety of genetic diseases. Abortion laws have been reformulated, all over the world, to permit, among other things, the termination of pregnancies for what are often called "fetal indications." Public opinion polls have revealed, over and over again, that the majority of people, even some who have their doubts about the wisdom of abortion for other indications, believe that it is better to terminate a pregnancy than to bring someone who is doomed to suffer with a serious disease or disability into the world. Hundreds of cost-benefit analyses have demonstrated that premarital and prenatal testing costs almost nothing when compared with the familial and social expense of caring for chronically ill and disabled people. Insurance schemes, both publicly funded and private ones, have agreed to pay, not only for the tests themselves, but also for the abortions that sometimes follow. Legislatures have passed laws regulating all these practices, thereby tacitly approving their fundamental goals. Dozens, perhaps hundreds, of entrepreneurs have developed profitable businesses supplying diagnostic equipment, each of them acting on the assumptions that people would keep wanting to have these tests performed, physicians would continue to offer them, legislatures would continue to encourage them, and third parties would continue to pay.

These developments have not been without controversy. Social support for the medical genetics project is by no means universal.

Objections come, for example, from some left-wing intellectuals, who see genetic testing as a manifestation of biological determinism, and as inescapably linked to the evils of the eugenics movement, including the outrages committed by the Nazis in the name of eugenics. Some of these intellectuals also regard genetic testing as a racist, perhaps even genocidal, enterprise, which seeks to restrict the reproduction of minority peoples.

Disability rights activists are also dubious about the value of genetic testing. They see it as a form of discrimination against the disabled, and they worry that a society that condones the abortion of fetuses that might be born blind or deaf or otherwise "differently abled" is not likely to accord full respect and rights to children and adults with those conditions.

Also opposed to genetic testing are people who oppose abortion in general and believe it should not be legal. From their perspective, all abortions are heinous crimes, but abortions performed by physicians (who have a professional obligation to care for the weak) in order to end the pregnancies of afflicted fetuses (the expected outcome when a genetic test has "positive" results) are the most heinous crimes of all.

Even some feminists agree with these other opponents of genetic testing. These feminists object to the medicalization of pregnancy, which, they believe, gives men too much control over women's bodies. They also note that being female is one of the "disabling" conditions that people test for and often, in some parts of the world, abort for. In addition, these feminists believe that the routinization of genetic testing diminishes women's autonomy rather than increasing it: a woman who fears she will incur social disapproval by resisting testing—getting off that gurney or refusing to sign that consent form—is a woman, this argument continues, who has become enslaved to the capitalist, patriarchal regime of quality-controlled reproduction. While they may not regard abortion for fetal indications quite the same way that pro-lifers do, these feminists agree that a society that encourages pregnant women to seek such abortions is a society with its priorities seriously out of whack.

This book had a very long gestation because of my own continuing ambivalence about the uses to which genetic testing has been put. As I read medical textbooks and medical journals, tracing citations backward, as I had been trained to do, I was also trying to decide whether I should have stayed on that gurney or run for the hills as soon as the consent form was proffered. As I read women's magazines and gathered samples of the informational brochures that are handed out in doctors' offices, I was also trying to decide whether prenatal diagnosis really did discriminate against the handicapped. As I interviewed medical geneticists and some of their patients, I was also trying to discover whether—as some of my feminist friends believed—the former were inherently evil and the latter inherently deluded. I was trying, in short, to resolve bioethical puzzles with historians' tools, by exploring the meaning of genetic testing to those who had developed it and those who had experienced it.

As the years wore on I began to realize two things. First, that many modern bioethical puzzles need to be resolved with historians' tools, because the puzzles are unprecedented and the usual tools—philosophical and theological—are not doing the job. Philosopher-bioethicists and theologian-bioethicists both seek and try to apply universal truths of ethical behavior: "Thou shalt not kill"; "Do unto others . . ."; "Above all, reduce suffering." Unfortunately, in the dilemmas that genetic testing and other new reproductive technologies create, these universal rules often conflict with one another. What, after all, is one to do if it turns out that, in this particular situation, the best way to reduce suffering is to terminate the pregnancy of an afflicted fetus? Historians have the insights that are needed to resolve such dilemmas, because the discipline of history teaches us to examine the particular, not the universal; we try, in other words, to uncover the very particular interests of very different people who are interacting in very particular situations. Once that is understood, we can begin thinking not only about the moral content of those interests but also about how they can be balanced so that no set of interests is allowed to dominate over the others.

The second lesson I learned from my research is that the motives of the people who developed genetic testing were motives that I could applaud. The search for bad genes—the mutated genes that cause devastating diseases and very disabling conditions—was carried out by good people who were acting ethically. The research medical geneticists who developed genetic tests, the clinical medical geneticists who administered them, the technicians who analyzed samples and provided test results, the counselors who helped patients make decisions, and most of all the patients themselves, the hundreds of women who subjected themselves to experimental procedures, the thousands upon thousands of women and men—"moral pioneers" as the anthropologist Rayna Rapp has called them—who subsequently had to decide, when positive results come back, what to do next; all of these people acted in the hope of reducing suffering, of making life less painful for people who carry, or who pass on, these disease mutations.

In this book, then, I make an ethical argument using historians' tools, exploring the history of genetic screening in order to demonstrate that most of its opponents are misguided. Since so many of these opponents regard medical genetics as an extension of the eugenics movement, I begin with a historical survey of that movement, not just in the United States and Germany (where the worst abuses in the name of eugenics occurred) but also in several other countries in which eugenicists sometimes succeeded in influencing public policy. Next I clarify the historical relationship among the eugenics movement, classical genetics, and medical genetics. I focus on classical genetics because it developed at the same time as the eugenics movement, and also because it is the science on which the practices of genetic testing are based. Molecular genetics, the genetics of DNA sequencing and protein formation, came later and did not alter these practices very much—at least not from the patients' point of view—although it vastly increased the number of diagnosable conditions.

Then I explore the complicated history of prenatal diagnosis, analyzing it as a sociotechnical system. Since the 1990s historians of technology have been examining technologies not as singular objects and not as networks of objects, but as sociotechnical systems, which link many objects with many social institutions. The computer on which I am writing these words is the paradigmatic example of such a system: it consists of several linked objects (CPU, printer, monitor, modem), which are linked to other objects (servers, for instance) through a variety of social institutions (electric companies, telephone companies, the instructional computing division of my university). This approach allows me to explain that, for example, changes in abortion law were as important in the diffusion of amniocentesis as changes in laboratory techniques and discoveries in classical genetics. I have also found this approach useful in developing a critique of some of the charges that feminists have made against medical genetics.

In the fifteen years between 1960 and 1975, three of the earliest genetic screening programs were implemented: newborn screening for phenylketonuria (PKU); carrier screening combined with fetal testing for Tay-Sachs disease; and carrier screening for sickle-cell anemia. (The terms "genetic testing" and "genetic screening" are often used interchangeably. Medical geneticists tend to use "genetic screening" when talking about large numbers of patients and "genetic testing" when the number of patients is small. Although this distinction is often fuzzy, and although it may sometimes be honored mostly in the breach, I will try to maintain it when I can do so without confusing the reader.) I narrate the history of the early screening programs by focusing on the interests of the individuals who developed the tests and lobbied for the programs. In the process I also address some of the critiques that have been made of these programs over the years. One chapter is wholly devoted to sickle-cell anemia screening programs in the United States in the early 1970s and the black communities for whom they were developed. An enormous amount of mis- and disinformation has been spread about screening for sickle-cell anemia, and the charges of genocide made by critics of these programs have left a legacy of mistrust between black patients and their physicians that persists to the present day.

The last of my substantive chapters concentrates on the only two mandated premarital genetic screening programs in the world: both of them on the island of Cyprus, both of them focused on the recessive gene that, when it is doubled, causes β-thalassemia. I went to Cyprus in the spring of 1999 to study these programs because they are mandated—and mandates are one of the aspects of eugenics that its modern opponents find most appalling. My experiences on Cyprus changed my mind about the wisdom of modern genetic testing; I went prepared to condemn and returned ready to applaud. On the island I discovered the existence of a moral consensus in favor of screening that unites physicians, patients, relatives of the patients, public health policymakers, and—most remarkable of all—the hierarchy of the Cypriot Orthodox Church, which is otherwise passionately opposed to abortion.

I conclude by discussing the critics of genetic testing in more detail and reprising the arguments made in the earlier chapters. My hope is to persuade opponents of medical genetics to rethink their positions and, more important, to provide some guidelines for all those people—doctors and patients, rich and poor, women and men, legislators, parents, business leaders, and citizens—who will have to make decisions about these complex matters in the years to come.

CHAPTER 1

Many Varieties of Beautiful Inheritance

In the early decades of the twentieth century eugenics was both a science, pursued by people who worked in laboratories, and a social movement, pursued by people who wanted governments to create new policies and new legislation. Today, many knowledgeable people regard the scientific part of eugenics as fraudulent, but no one denies that as a social movement it was, for a time, both popular and effective. Indeed, much of the current opposition to genetic testing and genetic screening depends on the memory of what eugenicists once believed in and the governmental policies that were motivated by those beliefs.

Some of the ideas on which the eugenics movement was based were ancient (Plato's *Republic* describes a more or less eugenic utopia), but most historians agree that eugenic ideas became particularly salient after the widespread acceptance of Darwinism, late in the nineteenth century. Francis Galton (1822–1911), a British gentleman-scientist, is widely regarded as the father of the modern eugenics movement. After graduating from Cambridge and coming into a modest fortune, Galton explored in Africa and the Near East. He also published some moderately important meteorological investigations. His life was transformed, however, in the early 1860s, when he read and became convinced by a book his first cousin had written. That book was *On the Origin of Species,* and its author was Charles Darwin.

Very soon after reading his cousin's book, Galton began to study the inheritance of human intelligence. His source materials were biographical dictionaries and his method was statistical. Galton used statistics to calculate the frequency with which eminent people were related to one another. He found, for example, that among a list of 391 artists, 65 were the father, the son, or the brother of another artist on the list, a frequency of 1 in 6.33. Among famous musicians the frequency was lower, 1 in 10, but remained, to Galton's mind at least, remarkable. So too, he thought, were the results for British statesmen (1 in 3) and "men of science and literature" (1 in 12). Galton's method was fraught with problems, but his conclusions were firm and unequivocal: talent and character must be hereditary. "If a twentieth part of the cost and pains were spent in measures for the improvement of the human race," he wrote, "what a galaxy of genius might we not create! We might introduce prophets and high priests of civilization into the world, as surely as we can propagate idiots by mating crétins."[1]

Over the next twenty years Galton pursued research on the inheritance of human traits. He sent questionnaires to the headmasters of schools; he asked physicians to send him family histories; at his own expense, he established a booth at the South Kensington Science Museum in which measurements could be taken of the parents and children who came to visit. His method remained statistical, or, to use his own term, anthropometric: he measured and counted and compared the data from one generation with the data from another. No matter what kind of characteristic he studied, Galton consistently determined that the influence of heredity was always more important than the influence of environment. In 1883 he published a book reporting on his researches: *Inquiries into Human Faculty.*

In this book Galton used the word "eugenics" for the first time;

he said he had made up the word by combining the Greek roots *eu,* meaning beautiful, and *gene,* meaning birth or inheritance, so as to convey the meaning "good in stock, hereditarily endowed with noble qualities." He went on to say that the word also denoted "the science of improving stock . . . which takes cognisance of all influences that tend in however remote a degree to give to the more suitable races or strains of blood a better chance of prevailing speedily over the less suitable than they otherwise would have had."[2] Thus, in Galton's initial formulation, eugenics was—as it continued to be long after his death—both a scientific research enterprise and a political program. The science of eugenics would reveal more about human traits and about how they are passed from one generation to another. The politics of eugenics would involve applying this knowledge in such a way that those who possessed the best traits would be able to outbreed everyone else. From the beginning, Darwin's theory of evolution by natural selection had been built into Galton's eugenic plans; if people with good traits, he believed, could only manage to outbreed people with bad traits, then the whole human stock would evolve in a positive direction, just as those animals and plants with the best adaptations go on to form new and improved species. In an article published in 1865 Galton had already expressed what would be the central idea of the politics of eugenics:

> The feeble nations of the world are necessarily giving way before the nobler varieties of mankind; and even the best of these, so far as we know them, seem unequal to their work. The average culture of mankind is become so much higher than it was, and the branches of knowledge and history so various and extended, that few are capable even of comprehending the exigencies of our modern civilization; much less of fulfilling them. We are living in a sort of intellectual anarchy, for the want of master minds.[3]

Galton, like many of his contemporaries, believed that the human species was composed of several subspecies that he sometimes called "nations" and other times called "races." He assumed, as many of his contemporaries also did, that in the struggle for human survival some races are more fit, or in his words, "more suitable," than others. A positive outcome for human evolution meant that the better races would have to prevail reproductively over the worse.

Eugenic Politics in Britain and the United States

Although Galton coined the term "eugenics" in the 1880s, a social movement based on his ideas did not become popular for another generation. Among his contemporaries, his political ideas were not very popular; the first enthusiastic eugenicists were people the age his children would have been, if he had had children. This generational delay means, ironically and significantly, that the word "eugenics" was coined three decades before the word "gene." In the early years of the movement, neither the scientific discipline called genetics nor the gene theory of inheritance had yet come into existence. Most of the early eugenicists were social reformers, not scientists; those who were scientists were called biologists (or, in a few cases, mathematicians). The early eugenicists worried about the frequency of what they thought of as traits (such as intelligence or susceptibility to disease) in a particular group of people, not about the frequency of a biological entity (a chromosome or a gene) in a particular population.

A British activist-socialite, Sybil Gotto, founded the Eugenics Education Society, based in London, in 1907. Gotto was, according to a friend and colleague, "a born organizer with an almost tireless energy which infected and stimulated all those who came in contact with her."[4] Galton, by then quite elderly and somewhat infirm, served as the first honorary president. The Society was one of several organizations fighting for social reform in Britain in the early decades of the twentieth century, all of them trying to solve what seemed to be the endemic problem of British pauperism. Although the members of the Society came from the political right as well as from the political left, they were all agreed on four propositions: (1) reform had to come from secular institutions, not religious ones; (2) the birthrate of the lowest classes in society (the "residuum" was the term the reformers often used for those classes) was higher than that of any of the so-called productive classes; (3) the residuum consisted, by and large, of people of very low intelligence; and (4) the solution to the problem of pauperism was to reduce the birthrate of these people, to zero if possible.

In Britain, then, the politics of eugenics focused more on class than on race (as defined either by skin color or by religion). The people whose marriage and birth patterns worried the reformers were unquestionably both white and British, although some eugenicists referred to the so-called residuum as if they belonged to a different subspecies from other Britons. In the early days, before modern Mendelian genetics had firmly taken root in British education, some eugenicists believed that environmental conditions (for example, excess consumption of alcohol or venereal infections) had caused the formation of this "degenerate" subspecies. The post-Mendelian position, which dominated after 1920, was that all conditions which produced pauperism (such as subnormal intelligence and alcoholism) were entirely hereditary and always had been. Either way, the eugenic goal remained the same: to stop paupers from reproducing.

British eugenicists championed five mechanisms to achieve this goal: education, segregation, reestablishment of workhouses, birth control, and sterilization. Segregation of the feebleminded (the term then used for people with below-normal intelligence) was their only substantial achievement—and it occurred early in the history of the movement. In 1913 Parliament passed the Mental Deficiency Act, which mandated that each school district should construct a special, residential, sex-segregated facility to which mentally deficient children could be remanded, for their entire lives if need be. Paupers, habitual drunkards, and unwed mothers on poor relief could also be incarcerated in these facilities. The eugenicists believed that feeblemindedness was probably a Mendelian recessive trait. If segregation by sex was maintained, they thought, so that people with below-normal intelligence could not mate (with one another or with others), then feeblemindedness—and with it pauperism—would die out within a few generations.

All the other eugenic efforts in Britain were either pyrrhic victories or failures. The educational campaign, to teach young people the facts of sex and the responsibilities of parenthood, was successful but not among paupers, and so defeated its own purpose. The birthrates of the "educated classes" and the "respectable poor" continued to fall, while that of "the lowest tenth" continued to rise. Eugenicists had tried to establish birth control clinics that would serve a clientele of pauper women, but they quickly failed, either for lack of financial backing or for lack of volunteer staff, or both.

Late in the 1920s the Society began to campaign for the legalization of sterilization, but by that time opposition to eugenics had become much stronger and more vocal. Opponents included Marxists (who regarded eugenicists as suffering from profound class bias), left-wing geneticists (who had figured out, by this time, that if mental deficiency were actually the result of a recessive gene, then sterilization of monozygotes would be ineffective, because carriers of the gene would still breed freely), and Catholics (who, following the dictates of the papal encyclical *Casti connubii* [1930], rejected any artificial interference with reproduction). In the mid-1930s, hoping to placate the opposition, the Society proposed legislation that would permit only voluntary sterilization, but even this was more than British legislators were prepared to accept at the time.

What the British were loath to consider, the Americans were quick to embrace. Like their British colleagues, American eugenicists were trying to solve two related problems: the economic burden of pauperism and the social burden of crime. They firmly believed, as an Indiana statute put it, that "heredity plays a most important part in the transmission of crime, idiocy, and imbecility."[5] They also believed, as the British did, that the proportion of degenerates (a term that included people who had committed crimes as well as those who were insane, alcoholic, or mentally deficient) in the population was increasing because degenerates had more children than "normal" Americans. Unlike the British, however, American eugenicists believed there was yet another cause for increasing degeneracy: European and Asian countries were sending degenerates to the United States as immigrants.

Thus American eugenic politics had two complementary goals: the first was to get as many "degenerates" (male or female) sterilized as possible; the second was to keep "degenerate immigrants" out of the country. By 1910 three states—Indiana, California, and Connecticut—had passed legislation permitting the involuntary sterilization of "confirmed criminals, idiots, imbeciles and rapists" who were confined in state-funded institutions such as prisons, insane asylums, and residences for the feebleminded. Involuntary eugenic sterilization was not without its critics, among them civil libertarians and Catholics, but in many states, for several decades, proponents outnumbered critics by a wide margin; eighteen more states had signed on to involuntary sterilization by 1940. In any event, many of the critics were essentially silenced when the constitutionality of the laws was upheld by the Supreme Court in the famous case *Buck v. Bell* (1927).

Carrie Buck was a presumably feebleminded young woman (test scores indicated that she had an IQ of about 50). She was the illegitimate daughter of an allegedly feebleminded mother who had been committed to the Virginia State Colony for Epileptics and Feeble-

Minded at the age of four. When she was in her teens, Carrie Buck gave birth to a daughter. The superintendent of the Colony, fearing that the infant too was feebleminded (he turned out to be wrong), decided that the mother ought to be sterilized before she could give birth again. Virginia had just passed an involuntary sterilization law, and to ensure that the law could pass constitutional muster the superintendent decided to make Carrie Buck's sterilization a test case.

On May 2, 1927, by a vote of eight to one, the Supreme Court of the United States decided that sterilizing Carrie Buck would not violate her constitutional rights. The *Buck v. Bell* decision allowed reproductive coercion, which might otherwise be repugnant, in the

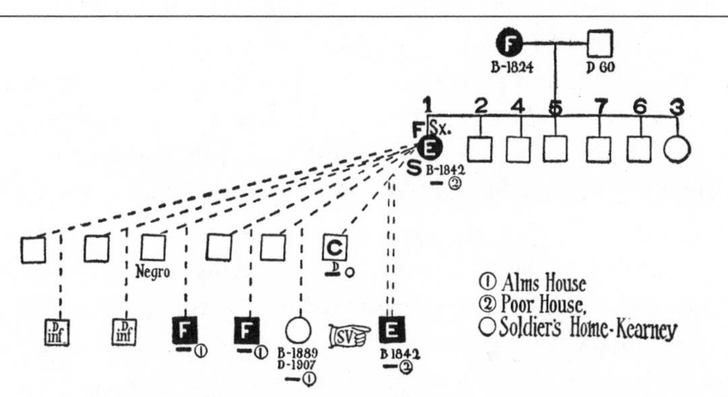

The "*poorhouse*" type of reproduction of the feeble-minded and epileptic. A lewd, feeble-minded and epileptic woman whose mother was certainly feeble-minded (but of whose father, brothers and sisters nothing is known) was the inmate of a county poorhouse. While there she had 6 children, of whom 2 died in infancy, 1 died at 18 in the almshouse, 2 were feeble-minded and are now living in the almshouse (1 the son of a negro) and 1 was epileptic, the son of a man with a criminal record. *C*, criminalistic; *D*, dead; *E*, epileptic; *F*, feeble-minded; *S*, syphilitic; *Sx*, sexually immoral.

The sort of pedigree that convinced legislators to vote for compulsory sterilization.

From Charles Benedict Davenport, *Heredity in Relation to Eugenics* (1911), 71.

interests of public health and safety: "It is better for all the world, if instead of waiting to execute degenerate offspring for crime, or to let them starve for their imbecility, society can prevent those who are manifestly unfit from continuing their kind. The principle that sustains compulsory vaccination is broad enough to cover cutting the Fallopian tubes."[6] By the middle of the 1960s, when the states began repealing (or ceasing to administer) these laws, roughly sixty thousand Americans had been surgically sterilized without their consent.

Spurred by the success that the early eugenics campaigners had had with state legislatures, eugenics societies began springing up in many American cities just before and after World War I. The most important of these was probably the Galton Society, which began meeting in New York in 1910, but there were others in places as different as Salt Lake City, Utah, and Chicago, Illinois. The American Eugenics Society, an effort to federalize these enterprises, was created in 1923 with almost thirty state-based branches. A research institute, the Eugenics Record Office, opened in 1910 and was funded by Mrs. E. H. Harriman, widow of a railroad magnate. Charles B. Davenport, the director of the Eugenics Record Office, was regarded by himself and others as the scientific leader of American eugenics. By 1914 universities as diverse as Harvard, Wisconsin, and the Utah Agricultural College were offering courses on eugenics. After 1920 eugenics education began appearing in high school curricula as well, sometimes in biology courses, sometimes as part of health education or home economics.

During those same years, eugenicists also succeeded in influencing American immigration policy. Some Americans had been worrying about the rising tide of immigration since the 1840s. They had noticed that some immigrants arrived poor and remained poor, constituting a disproportionate number of those who broke the law, required charity, or resided in caretaking institutions funded by the state. Even before there were eugenicists, some laws restricting immigration had been justified on the grounds that the new arrivals either were of "defective stock" or were a threat to American "racial integrity." After the creation of an organized eugenics movement, these arguments gained both currency and strength because the eugenicists were able to argue that the phenomenon had a biological basis.

Harry H. Laughlin, who was second in command at the Eugenics Record Office, became the star witness when the United States Congress considered what would eventually become the Immigration Restriction Act of 1924, the law that virtually closed the door to immigrants who were not Protestants from northern Europe. Albert Johnson, majority chairman of the House Committee on Immigration and Naturalization and a member of the American Eugenics Society, had invited Laughlin to testify. Laughlin knew how to make his eugenic arguments persuasive. He told the committee that he had done scientific studies of the inmates in institutions for the insane and the feebleminded. He had found, he said, that those institutions housed a disproportionate number of immigrants from Eastern and Southern Europe; they must, he thought, have been afflicted with hereditary degeneracy before they arrived. Among the charts and tables he presented to the committee was a compendium of the most unsavory photographs from Ellis Island, which he entitled "Carriers of the Germ Plasm of the Future American Population." Laughlin's strategy worked: the 1924 law created quotas, very low ones, for immigration from the countries he had identified as the ones responsible for exporting degeneracy.

Although involuntary sterilization and immigration restriction were the two main planks of eugenic platforms in the United States, there were other eugenic measures that were ratified by state or federal action. Numerous state laws prohibiting miscegenation (sexual relations and marriages between persons of different races) were justified on the grounds that "racial hybridization" would lead to degeneracy. In many states marriages between "unfit" persons were prohibited altogether. In Connecticut, for example, the penalty for violating the law (that is, the penalty for the person who performed a marriage ceremony in which one or both of the parties was feebleminded) was three years' imprisonment. In Indiana the feebleminded, those with a venereal disease, and habitual drunkards were not allowed to marry. Finally, in the interwar years, many of those who campaigned successfully for relaxation of the state and federal laws prohibiting the distribution of birth control information and devices were eugenicists; they hoped that if poor women were provided with diaphragms the number of criminals, degenerates, and paupers would diminish in succeeding generations.

Thus, although some people think of Germany as the place in which eugenics flourished and others think of Britain as the place in which it began, when measured by the number and longevity of eugenic programs, eugenics was most successful in the United States.

Eugenic Politics in Catholic Countries

Political movements can mutate, just as genes can. In both Britain and the United States, Catholics were opponents of eugenics, but in countries dominated by the church there were still active eugenics movements, albeit of a very different sort. In several Catholic countries—France, Brazil, Argentina, and Mexico—eugenics movements tended to focus on positive (encouraging the fittest to breed the most) rather than negative policies. In addition, Catholic eugenicists continued to believe in the inheritance of acquired characters long after their Protestant colleagues had adopted the idea that "germplasm" was continuous and could not be affected, in any significant way, by the environment.

At the turn of the twentieth century the French, like the British and the Americans, were worried about the physical and mental degeneration of their population, but they were also concerned about another significant decline: the size of the French population had been falling since the middle of the nineteenth century. Not surprisingly, there were calls for "biological regeneration" after France was defeated by Germany in the Franco-Prussian War of 1871 and, again, after thousands of young Frenchmen died in the trenches of World War I. French eugenicists, most of whom were either physicians or anthropologists, organized themselves as the French Eugenics Society in 1912 and managed to publish a periodical, *Eugénique,* which ran until 1926. The biological basis for the reform measures they advocated was both Lamarckian (in the early nineteenth century the French zoologist Jean-Baptiste Lamarck had argued that acquired characters were often inherited) and pronatalist.

Birth control was anathema to the French eugenicists of the early twentieth century. So was involuntary sterilization, which according to one spokesman was "repugnant to our need for liberty and our delicate individualism."[7] Instead, the French eugenics movement advocated a variety of measures that would encourage Frenchwomen to have more and healthier children. As Lamarckians they believed that qualities acquired during a lifetime could be passed on to the next generation. They wanted "eugenical" policies that would foster the health of mothers and children, including health examinations before marriage (to ensure that the partners carried no transmissible diseases); public health programs to control tuberculosis, venereal disease, and alcoholism; construction of more sanitary housing, with space and priority for families with numerous children; and financial benefits for parents with more than one child. Because the eugenicists allied themselves with other advocacy groups, some of these provisions became law. The National Office of Social Hygiene was established in 1924 to create the public health programs; construction on government-funded housing for large families also began in the 1920s; the premarital health exam was made mandatory under the Vichy government in 1942; and paid maternity leave and other child care benefits were features (still in existence) of postwar reconstruction policy.

In several Central and South American countries eugenics movements followed the French model—with a remarkable racial twist. In each of these countries (Brazil, Argentina, Mexico) the earliest eugenicists were secularists, wanting to base social reform on modern science; they were also members of a professional elite, largely physicians and university faculty members. Like their French counterparts, they were also Lamarckians, convinced that improvements in public health and sanitation would—over the course of a few generations—improve the biological and mental character of the population. Unlike their French counterparts, however, these Spanish- and Portuguese-speaking eugenicists were themselves *mestizos;* like all members of Central and South American elites, their families were derived from the interbreeding, at some time in their history, of European colonizers with indigenous Indians and enslaved Africans. European eugenicists and anthropologists liked to point to the interracial breeding patterns of Central and South America as examples of bad eugenic practice, the production of degenerate hybrid races; they believed that the best features of the "whites" had been diluted by the importation of "black" and "brown" bloodlines.

Needless to say, many Brazilian, Argentinean, and Mexican eugenicists, especially the nationalists among them, could not abide a set of notions that suggested that neither they nor their countrymen were as fit as Europeans. So they turned the argument on its head, claiming that hybrid races were actually more vigorous and adaptable than "pure" races. Indeed, they argued, white bloodlines added superior mental characteristics to the fundamentally good health (particularly in tropical climates) of the black and brown races. If infant mortality was higher in the Americas than in Europe, if alcoholism was more rampant, if families were smaller, it must be because sanitary conditions were worse, not because heredity had doomed the children of mixed-race marriages. Although there were a few significant exceptions, particularly in the first generation (before 1920), the majority of Latin eugenicists were not racists in anything akin to the sense in which many North American and North European eugenicists were. Northerners abhorred miscegenation while southerners embraced it.

Left-Wing and Soviet Eugenics

In both North and South America, in countries dominated by both Protestant and Catholic citizens, most—though not all—eugenicists were political conservatives. In the United States and Britain a few socialists and feminists became involved in eugenic politics; access to birth control, they thought, would improve the lives of the poor. In both countries there were also some Marxist scientists who became eugenicists; if freed of its class and race biases, eugenics, they believed, could be combined with communism to build a better future for everyone. In the Soviet Union, in contrast, left-wing biologists found themselves trying to convince communist officials of the social value of eugenics so that, in a period of turmoil and scarcity, funding for their research institutes would be continued.

Hermann J. Muller, the American geneticist who won the Nobel Prize for his discovery that radiation increases the frequency of mutation, was an outspoken socialist and eugenicist. In the late 1920s Muller ran into trouble with the trustees of the University of Texas, where he was then teaching, because he refused to recant his radical views. By 1933 he had fled to the Soviet Union to become the director of the genetics laboratory of the Leningrad Institute of Applied Botany. J. B. S. Haldane, the prominent British biologist, was a Marxist who criticized British proponents of involuntary sterilization as being both inadequately eugenic (because sterilization would not actually have the eugenic result they predicted) and classist (because they assumed that the middle class was more fit than the working class). Muller, Haldane, and other left-wing eugenicists took up Galton's original notion that eugenics was a scientific moral creed, a fitting replacement for Christianity in industrial societies. They were all Mendelians, which meant that they regarded the public health measures advocated by the French and Latin eugenicists as ineffective from one generation to the next. In place of sterilization, segregation, and immigration restriction, these men of the left advocated what Muller came to call "germinal choice," or what we now know as either artificial insemination or *in vitro* fertilization. Sex should be separated from procreation, they argued, and individual women should be educated to choose sperm from "the best" men, or, if education did not work, the state should require them to make such a choice.

For Soviet eugenicists, struggling first with the privations of the 1920s and then with the ideological wars of the 1930s, germinal choice turned out to be a luxury they could not afford to advocate. Even before the tsar was overthrown there had been some interest in eugenics in Russia. Galton's first book, *Hereditary Genius,* published in English in 1869, was translated into Russian six years later; Davenport's *Heredity in Relation to Eugenics,* which first appeared in 1911, was translated in 1913. As the revolutionary period drew to a close and an institutional structure for Soviet science began to emerge, two Russian biologists, Nikolai Konstanovich Koltsov and Iurii Aleksandrovich Filipchenko, seized the opportunity to have a Bureau of Eugenics established as part of the newly created Commission on the Study of Natural Productive Forces. A year earlier, in 1920, they had organized the Russian Eugenics Society. Between the Bureau, which funded research, and the Society, which published a journal, *Russkii Evgenischeskii Zhurnal,* both Soviet genetics and Soviet eugenics seemed off to a healthy start.

The early Soviet eugenicists believed that the new biology would provide a "civic religion" appropriate to socialism, but although they had no difficulty establishing a research agenda, a practical political program that would be both acceptable to the new government and compatible with their genetic findings was harder to achieve. In the wake of the Russian Revolution, the size of the population was collapsing: between 1917 and 1920 Moscow lost almost half its people and Petrograd lost roughly 70 percent. All over the country, deaths exceeded births by large margins: 243 per 10,000 in Moscow, 484 in Petrograd, and probably even more in the countryside. Sterilization of any citizens, even those clearly "unfit," was absolutely out of the question. In 1929, however, Aleksandr Serebrovskii, one of Koltsov's students, proposed that there was a biologically sensible solution to this demographic problem. Writing in response to a call to Party members to discuss a new Five Year Plan, Serebrovskii recommended "the widespread induction of conception by means of artificial insemination using recommended sperm, and not at all necessarily from a beloved spouse." Serebrovskii went on, in terms he must have hoped would appeal to the Party's Central Committee:

> With the current state of artificial insemination technology . . . one talented and valuable producer could have up to 1,000 children . . . In these conditions, human selection would make gigantic leaps forward. And various women and whole communes would then be proud . . . of their successes and achievements in this undoubtedly most astonishing field—the production of new forms of human beings.[8]

Unfortunately for Serebrovskii and, indeed, the whole of the Soviet genetics establishment, neither the grandiosity of his proposal for germinal choice nor its dependence on Mendelian concepts appealed to Stalin.

Sometime in the early 1930s a minor agricultural official, Trofim Lysenko, had convinced Stalin that Mendelism was bad biology and bad politics: bad biology because it contradicted Lysenko's own convictions about heredity, which were Lamarckian; bad politics because it contradicted Marxism when it suggested that human characteristics were determined biologically, not economically. From Lysenko's perspective, the Soviet geneticists who advocated Mendelism and eugenics were bourgeois enemies of the state—and Stalin had a very particular set of plans for bourgeois enemies of the state. By 1940 all the Mendelian geneticists in the Soviet Union were dead, imprisoned, or living in exile. Soviet eugenics had ceased to exist. And so had left-wing eugenics in the West. In the 1950s, as the eugenics of the Nazis became notorious, as the totalitarian character of Soviet socialism became apparent even to the truest of the true believers, and as the exigencies of the Cold War began, all those British and American Marxists who had once advocated either communism or germinal choice or both somehow came to the same conclusion: the less said about their earlier political convictions, the better.

Eugenics under the Nazis

German eugenics originated in the final decade of the nineteenth century, and its history might have been indistinguishable from American or British eugenics had it not been for Hitler. Even before coming to power Hitler was both a eugenicist and a virulent racist. Between 1933 and 1945 he put these prejudicial ideas into dictatorial practice with devastating consequences.

Those consequences were not intended or anticipated by the founders of German eugenics, who were not racists, at least not of the Hitlerian sort. Many early German eugenicists were discouraged physicians. The kind of medicine they had been taught to practice was not doing much good as tuberculosis raged unchecked, infant mortality remained high, and alcoholism and venereal disease were, they thought, everywhere fraying the social fabric. As advocates of social Darwinism, early-twentieth-century German eugenicists believed that medical care was actually diminishing the level of public health by sustaining the weak so they could live long enough to reproduce. In the minds of these physicians, a new kind of medicine was needed. They called for a medicine based not on scientific physiology but on scientific genetics, a medicine that would concern itself as much with the welfare of the race as with the welfare of individuals. The term they created for this new medicine was "Rassenhygiene," meaning racial hygiene. By "race" they seemed to have sometimes meant "the German people as a whole" and other times meant "the human race as a whole."

Neither anti-Semitism nor Nordic supremacy was part of the original program of *Rassenhygiene.* Alfred Ploetz, the physician who coined the term in 1895, believed that Jews and Aryans were among the most cultured of the world's peoples, that there was no such thing as a "pure" race, and that racial hybridization was actually "a means of increasing fitness and as a source of good variations."[9] Wilhelm Schallmeyer, another physician-founder of German eugenics, preferred the term "Rassehygiene"; without the *n* following *rasse,* the word becomes singular, not plural, making it very clear that Schallmeyer was concerned with the entire population of human beings, no matter what their skin color or religion. Yet another eugenicist, Alfred Grotjahn (known today as one of the architects of German social welfare legislation), preferred the Germanized version of Galton's original word—"Eugenik"—so that there would be no confusion with racism at all.

These early German eugenicists were not quite sure what social programs they wanted most to advocate. Genetics should surely be taught in all the realms of higher education, especially medical education. Doctors should learn to detect hereditary conditions—particularly feeblemindedness, insanity, and alcoholism—and to counsel their afflicted patients to refrain from parenthood. Some thought that sterilization might be most successful. Still others felt that segregation (on the English model) would be more effective.

Birth control was surely a very bad idea, they argued, and so was feminism; healthy women should be having as many children as physiologically possible and should be devoting their full attention to the well-being of those children. Venereal disease should certainly be controlled because of its negative effects on children born to infected mothers. Perhaps the government should include family allowances in its schedule of welfare benefits—but only, of course, to the families of those who were "fit." Perhaps there should also be compulsory marital counseling, so that those who had made dysgenic marriages, unions likely to yield "unfit" offspring, could be encouraged not to reproduce.

Above all, the early German eugenicists believed that more research was needed. In the first decades of the twentieth century Germany had the best-funded scientific research establishment in the world. The eugenicists wanted funding for genetics, especially human genetics. They worked hard to achieve this goal, and their work was rewarded. The *Archiv für Rassen- und Gesellschaftsbiologie* (Archive for racial and social biology), founded by Ploetz in 1904, was the world's first scientific periodical specifically devoted to publishing research in what would now be called human genetics. By the time Hitler rose to power in 1933, there were twelve other scientific periodicals devoted to the same subject, and German geneticists were regarded as international leaders in the field. Fritz Lenz, the physician who succeeded Ploetz as the editor of the *Archiv,* was appointed to the nation's first professorship of racial hygiene, at the University of Munich, in 1923. A government-sponsored research institute, the Kaiser Wilhelm Institute for Anthropology, Human Genetics, and Eugenics, was created in 1927. Its first director was Eugen Fischer, a physician turned physical anthropologist. A few years earlier Fischer and Lenz had coauthored (with Erwin Baur, another eugenicist) one of the earliest textbooks in the field of human genetics, *Grundriss der menschlichen Erblichkeitslehre und Rassenhygiene* (Foundations of human genetics and racial hygiene). This was very quickly translated into English under the title *Human Heredity* and was widely praised by geneticists of many political persuasions all over the world. It is still regarded as one of the foundational texts of the discipline of human genetics.

After its traumatic defeat in World War I, Germany endured more than a decade of internal disarray and external humiliation. Domestic political turmoil was coupled with hyperinflation caused, in part, by the reparations Germany was forced to pay to the countries that had been devastated by the war. Under such conditions, what had once been moderate political reform movements—such as the eugenics movement—became both polarized and radicalized. Ploetz's views about racial purity and about Jews began to shift, Lenz became more insistent on the pervasiveness of hereditary insanity and the wisdom of compulsory sterilization, and the words "Nordic race" and "Nordic supremacy" began to appear more frequently in eugenics publications. Moderate eugenicists became so disturbed by the radicalization of their movement that they seceded from the Society for Racial Hygiene and created their own organization, *Deutscher Bund für Volksaufartung und Erbkunde* (The German association for ethnic improvement and genetics). They declared, in the first issue of their journal, published in 1926, that they wished to be "free of all political and religious positions and free from all particular racial tendencies."[10]

By that time, however, moderate eugenics was probably doomed in Germany. Hitler had already begun to advocate the racist version of eugenics, and so had his political party, the National Socialists. In truth, "advocate" may be too weak a word, for racist eugenics was a cornerstone of Nazism. Some Nazis argued that National Socialism was nothing more than applied biology or, closer to the mark, applied genetics. Writing in the party's magazine in 1930, one Nazi pointed out that this was precisely what distinguished National Socialism from Marxist socialism. Whereas the Marxists assumed that people were biologically equal, the Nazis knew that they were not. The author of the article, Theobald Lang, articulated precisely the terms on which Nazi eugenics differed from its liberal German and its Marxist antecedents:

1. The genetics of particular individuals, races, and race mixtures are different.
2. We cannot consciously change the traits specified by genetics. . . ; [and]
3. The current [liberal, Weimar] economic order and conception of civilization exert a negative selection on future generations, having as a consequence a general decimation of German peoples and thereby the entire world.[11]

Echoing these points, a Nazi physician declared: "National Socialism without a scientific knowledge of genetics is like a house without an important part of its foundation."[12] Hitler agreed and thought that doctors would be more important to the success of his regime than lawyers, engineers, or manufacturers. He later referred to the Third Reich as "the final step in the overcoming of historicism and the recognition of purely biological values."[13]

Within weeks of assuming power in 1933, the Nazis began retooling virtually every political and social mechanism at their command to achieve eugenic goals. On July 14, 1933, the government promulgated the Law for the Prevention of Genetically Diseased Offspring, otherwise known as the Sterilization Law. Until that day sterilization had been illegal in Germany. A whole new system of genetic health courts was created. Physicians were required, on pain of a substantial fine, to report to their local genetic health court any patient suffering from feeblemindedness, schizophrenia, manic depression, deafness, epilepsy, Huntington's disease, congenital (from birth) blindness, or severe alcoholism. After investigating each case, the local court could, sometimes secretly, mandate that the individual be sterilized. There was a central genetic court in Berlin, presided over by lawyers and doctors, to which appeals could be directed. Biologists and geneticists were to be paid by the court system to present expert testimony about health pedigrees at the trials. The courts also paid physicians to perform the surgical sterilization.

The genetic health courts remained in business throughout the war years (1939–1945), but most of the estimated four hundred thousand sterilizations that they ordered occurred before 1939. Some historians believe that virtually every adult in Germany who suffered from one of the listed afflictions had already been sterilized by that time. Others argue that the cost, both in money and in manpower, of sustaining the sterilization program turned out to be too great as Germany prepared for and then engaged in war. Whatever the reason, there is no doubt that in the summer of 1939 the Nazis stopped sterilizing disabled people and began murdering them.

This "euthanasia" program began with crippled children. An agency created by Hitler in the summer of 1939, the Committee for the Scientific Treatment of Severe, Genetically Determined Illnesses, requested all state governments to have physicians and midwives register the birth of any child born with a congenital deformity. This included "idiocy or Mongolism (especially if associated with blindness or deafness); microcephaly or hydrocephaly of a severe or progressive nature; deformities of any kind, especially missing limbs, malformations of the head, or spina bifida; or crippling deformities such as spastics."[14] When these registers of "lives not worth living" were finished (some states never collected the data), they were sent to Berlin, where officials determined—solely from the medical records—which infants and children were to be "selected" for admission to one of the twenty-eight hospitals equipped with extermination facilities. The children were killed by carefully chosen methods such as the slow administration of poison which would not leave telltale marks, so that parents could be given false explanations for the death. They might be told their child had died of appendicitis or brain edema. Parents were also told that, because of the fear of epidemics, their child's body had been immediately cremated. Altogether, roughly five thousand children were killed this way.

Only a few months after first targeting children, the Nazis turned on mentally ill adults who were institutionalized. The physicians and bureaucrats who planned the program estimated that out of every thousand Germans, ten needed some form of psychiatric care; five of those needed to be institutionalized for long periods of time; and one out of those five should be destroyed, so as not to be a continuous drain on the resources of the Reich. Quotas were established. Mental hospitals were outfitted with gassing chambers disguised as showers and with crematoria for disposal of the bodies. The directors of the hospitals were instructed, but not required, to apply for permission to kill their patients. These applications were vetted by central committees of physicians, who were themselves instructed to grant permissions in accordance with the quotas. Within two years the national quota had been met; roughly seventy thousand mental patients, some of them simply elderly victims of dementia, had been gassed. As complaints, particularly from the Catholic church, began to mount, Hitler stopped the program and decentralized it, giving individual hospital directors permission to gas patients if they needed to. Some continued murdering patients until they were forcibly removed from their directorships in the first months of the postwar occupation of Germany.

During the decade between 1935 and 1945, marriage and sexual relations were also regulated by Nazi eugenic goals, the desire to keep the Aryan race both "pure" and "healthy." In the autumn of 1935 Hitler signed both the Law for the Protection of German Blood and German Honor and the Law for the Protection of the Genetic Health of the German People. (These two, plus the Reich Citizenship Law, are generally referred to as the Nuremberg Laws.) "Protecting German blood and honor" meant prohibiting marriages and sexual relations between Aryans and non-Aryans. "Protecting the genetic health of the German people" meant prohibiting marriages between those who were genetically "sick" and those who were genetically "fit." People who were genetically "sick" could marry each other, but only if they first consented to sterilization. Implicit in both laws was the notion that non-Aryans—principally Jews, but also Slavs and gypsies—were genetically inferior in exactly the same way as Aryans who had congenital or sexual diseases. The offspring of these couplings, by definition, would be "unfit" and inferior, thereby diminishing both the health and the quality of the German population.

The bureaucracy needed to enforce these laws was simply astonishing. Since no one could obtain a license to marry without first getting certification and counseling from one of the health offices (742 of which had been established by 1937), thousands of trained people, including physicians, nurses, and technicians as well as thousands of clerks, were employed in the effort to evaluate and file genealogies. The rules were exceptionally precise: "Individuals with only one-quarter Jewish ancestry were considered Germans and allowed to marry other Germans—unless those other Germans were themselves one-quarter Jewish, in which case there was the danger that some of the offspring would be half Jewish (according to a peculiar kind of Mendelian logic), and so the marriage would be illegal."[15] All over the country genetics research units were created, employing hundreds of additional biologists and clerks, just for the purpose of helping individuals compile these genealogies.

During the Nazi period eugenics education was added to the curriculum at all levels of schooling. Biology textbooks were altered to include material not just on genetics but also on race. High school mathematics students were asked to solve problems such as these:

> In one region of the German Reich there are 4,400 mentally ill in state institutions, 4,500 receiving state support, 1,600 in local hospi-

> tals, 200 in homes for the epileptic, and 1,500 in welfare homes. The state pays a minimum of 10 million RM/year for these institutions. What is the average cost to the state per inhabitant per year?
>
> The construction of an insane asylum requires 6 million RM. How many housing units @ 15,000 RM could be built for the amount spent on insane asylums?[16]

People training to become teachers were taught physical anthropology so that they could teach their students to distinguish between Aryans and members of other races. Doctors wishing to enter military service were required to attend special training camps in which they were taught the fundamentals of racial science and the need for preventive racial hygiene. Vast quantities of posters and informative pamphlets were published, all of them illustrated with charts, drawings, and cartoons—so that even the barely literate would get the message.

Apparently the message fell on willing ears. How else to explain the participation and tacit consent of so many Germans in what turned out to be the ultimate eugenic measure of them all: the segregation in ghettos and concentration camps and subsequent murder of six million "enemies of the Reich," men, women, and children—most, but not all of them Jewish—whose only crime was having lost the toss of the genetic coin, having been born into one of the groups that were thought to be destroying the genetic health, the racial purity, the biological superiority of the German people.

Thus under Nazi eugenic programs six million people were categorized as members of degenerate races and were slaughtered; four hundred thousand "pure-blooded" Germans were categorized as genetically defective and were sterilized; and seventy-five thousand disabled "Aryans" whose care was deemed "too expensive" were murdered in hospitals, sanatoria, and nursing homes. Nazi eugenics has often been called genocidal, but in a bizarre way, that is too limited a term, for the Nazis turned on their "own" as well as upon the so-called aliens who lived among them. The Nazis were also uniquely compulsive and uniquely murderous eugenicists: American and British eugenicists might want to sterilize institutionalized people, but Nazi eugenicists wanted to sterilize anyone—neighbor, relative, friend—who was "genetically" abnormal. American and British eugenicists were willing to stop at sterilization to improve the national breeding stock; Nazi eugenicists were willing to proceed to murder.

Eugenics in Scandinavia

Had Germany not been ruled by a dictator who was an obsessive racist, the nation might have adopted eugenic laws similar to the ones passed in the Scandinavian countries during the interwar years. In the first three decades of the twentieth century, the eugenics movements in Norway, Sweden, Denmark, and Finland were very similar to those in Germany. Scandinavian eugenic reformers, just like their German colleagues, were concerned about the size and the quality of their populations. They were convinced that both were deteriorating because many of Scandinavia's "best" people had emigrated. They coupled these concerns with Darwinian social attitudes and Mendelian genetic convictions to create a eugenics movement.

Scandinavian eugenicists began to lobby for sterilization laws in the early 1920s, but there was considerable resistance and, since all the countries were parliamentary democracies, very little action for more than a decade. However, Danish eugenicists finally succeeded in getting a sterilization law passed in the waning days of 1929. Norway, Sweden, and Finland rather quickly fell into line in 1934 and 1935.

The Scandinavian sterilization laws were quite different from those in Germany. First, they were directed at a smaller and more specific population: those who were so mentally retarded as to be unable to function in society. The laws were not applied to people who were alcoholic, physically crippled, mildly retarded, epileptic, blind, or deaf. Second, the laws provided only for voluntary (or quasi-voluntary) sterilization. Individuals, or their legal guardians, had to initiate requests and had to consent to the operation. In some countries, most notably Sweden, mechanisms were put into place to "invite" consent, but if the individual continued to resist, the operation could not be legally performed. No agency of the government could mandate sterilization. The measures were not implemented in a racist fashion. The ethnicity of individuals was not at issue, only their mental capacity.

In the interwar period all the Scandinavian countries were parliamentary democracies. As a consequence, eugenic laws were passed only when the eugenicists could ally themselves with other reformers, specifically those who were interested in protecting the welfare of women and children. Many advocates of sterilization of the mentally incompetent argued that it was a far more humane way to prevent them from bearing children than segregating them in institutions, as the British and Americans sometimes did. Others argued that it was important that children not be born to people who were incapable of rearing them properly. The wording that eugenicists proposed for the Norwegian statute makes this collaboration between them and other reformers very clear:

> Persons who are insane or are mentally defective can be subjected to sexual surgery at the request of a guardian, when there is reason to assume that the person will not be able to provide for the needs of self or offspring, or that a mental disease or grave physical defect will be passed on to the offspring, or that because of abnormal sexual instinct he is likely to commit sexual crimes.[17]

Thus the Scandinavian sterilization laws of the interwar years were understood, both in Scandinavia and elsewhere, as part of a package of social welfare reforms. Eugenic abortion, a provision that the Nazis never countenanced, was also authorized in most of these interwar laws; abortion was permitted when physicians could demonstrate the likelihood that a pregnancy was likely to result in a seriously ill or disabled child, a child that would threaten the welfare of its mother and the other people who were dependent on her.

The Scandinavian eugenic laws have been modified in the decades since they were instituted, but they have never been completely repealed. Although there has been vocal, local criticism of each of them, most Scandinavian legislators seem to have agreed that when sterilization and abortion are voluntary, rather than mandated, and when the ultimate goal is improvement of the welfare of individuals, eugenic actions should be permitted. The authors of the European Union's 1997 Convention on Biology and Medicine may have had precisely these laws in mind when they decided that some eugenic measures might be ethically acceptable if they "are necessary in a democratic society in the interest of public safety, for the prevention of crime, for the protection of public health or for the protection of the rights and freedoms of others."[18]

In political terms, eugenics was a very mixed bag. Some eugenicists were very far to the political right, while others were just as far to the left—and a fair number of others could be located in the moderate middle. Some eugenicists were absolutist hereditarians, but others were as focused on improving the health of mothers and babies as on improving their germplasm. The impulse to improve the human race by improving its genetic heritage sometimes led to morally egregious programs (such as involuntary sterilization in the United States and Nazi Germany), sometimes to morally commendable programs (such as family benefits in Scandinavia and France), and sometimes to programs on which moral thinkers have disagreed (such as voluntary sterilization in Scandinavia). In some places and at some times, eugenic impulses motivated research that has enhanced our ability to care for people with genetic diseases. In other places and at other times, eugenic impulses motivated research that led governments to kill or maim people with genetic diseases. Some programs that some eugenicists favored (such as free access to birth control information and technologies) are now approved of by millions of other people; some programs that other eugenicists favored (such as germinal choice) are roundly scorned.

The word "eugenics" is one of those nouns to which people often have knee-jerk reactions. Call something "eugenic" and most people will immediately think "Nazi" and be repulsed. After 1945 many eugenicists were afraid of precisely this reaction. The American journal *Eugenics Review* became the *Journal of Human Genetics*. In Britain, *Annals of Eugenics* became *Annals of Human Genetics*. The Biological Laboratory at Cold Spring Harbor sandblasted the lintel of the building that had once been called the Eugenics Record Office. Lionel Penrose, holder of the Galton Eugenics Chair at University College, London, "found it a 'continual embarrassment' to have to explain that both his laboratory and the professorial chair were 'wrongly named.'" He eventually persuaded the college authorities to rename his professorship the Galton Chair in Human Genetics.[19]

In the immediate post-Nazi era such stereotyping of eugenics was perhaps understandable, but by the end of the century many thoughtful people had concluded that some eugenic dreams were morally right even if others had been ethical nightmares. With the passage of time, historical and comparative studies of eugenics have demonstrated that—like all stereotyping—wholly negative characterizations of eugenics are unfair, unjust, untruthful, and a very bad basis for policy decisions.

CHAPTER 2

Eugenics and the Genealogical Fallacy

Late in the winter of 1901 a British physician, Archibald Garrod, began an exchange of letters with a British biologist, William Bateson. Although neither man could have known it at the time, those letters—plus the two articles and a book which they published within the year—represent the origin of the discipline that is today called medical genetics. Garrod and Bateson had a lot on their minds during the winter and spring of 1901; each was trying to solve a puzzle to which the other held a crucial clue.[1]

Medical geneticists are the people who develop, administer, and evaluate genetic tests. As its name suggests, medical genetics combines medicine—the study, diagnosis, prevention, and treatment of disease—with genetics—the effort to understand how traits are passed from one generation to the next. Although medical genetics was not named until the 1930s, its history began in that early barrage of letters between Garrod, then living in London, and Bateson, who resided eighty miles away, just outside Cambridge.

Archibald Garrod was born in 1858. His father, Alfred Baring Garrod, was a physician and a researcher, a disciple of the French experimentalist Claude Bernard. In 1865 Bernard had set the medical world on its ear by publishing a call to chemical arms, *An Intro-*

duction to the Study of Experimental Medicine. Bernard argued that the body was a chemical system, and that diseases could not be properly treated until their underlying chemical causes were understood. Alfred Garrod took up Bernard's challenge; through careful, quantitative biochemical studies, he discovered that patients with gout had increased concentrations of uric acid in their blood, a crucial discovery that made him a medical celebrity. Young Archibald was determined to follow in his father's footsteps. Within months after finishing his medical studies, he was already publishing the results of his research: first, very simple reports of interesting medical cases; then, in 1892, his first venture into chemistry, analyses of the chemical changes in the blood of patients with rheumatoid arthritis. Between 1893 and 1898, sometimes with the help of his lifelong friend the chemist F. Gowland Hopkins, he did a series of careful chemical studies of the pigments in urine.[2] In those last years of the nineteenth century, physicians could not be employed as researchers; if they wanted to do research they had to squeeze it into whatever time they had left over from seeing patients.

In 1898, while he was experimenting with urinary pigments, Archibald Garrod was introduced to a patient with alkaptonuria, black urine. Physicians had known for much of the previous decade that the chemical that turned urine black was homogentisic acid. By experimenting with the urine of this particular patient, Garrod developed a simple method for extracting the acid from the urine so that its concentration could be measured more efficiently.[3] Had his researches stopped there, Garrod would be little known today; black urine is a bit scary to the patient, but it is otherwise harmless, which means that measuring the concentration of this particular urinary pigment has almost no therapeutic value. But Garrod had discovered that some of the brothers and sisters of the original patient also had black urine; his decision to test the urine of all the patient's relatives set him on a very different research track, a track that ended up making his work both unique and notable.

The research reports on alkaptonuria that Garrod published between 1898 and 1902 read like detective stories. He discovered that although some of the patient's brothers and sisters had black urine, neither of his parents did. Could the "normal" members of the family just have mild cases of the disease? No, Garrod learned, there is no such thing as "gray" urine; in afflicted families, alkaptonurics have a lot of homogentisic acid in their urine (they secrete two to six grams every hour), but normal family members have none at all. "Its appearance in traces, or in gradually increasing or diminishing quantities, has never yet been observed."[4]

Widening the investigation to other afflicted families made no difference; the results were always the same. Could the condition stem from some kind of chronic urinary tract infection? Not likely: no trace of bacteria could be found in the urine of alkaptonurics, and many reported that their urine had been black from the day they were born. "In one of my cases," Garrod wrote, "the staining of the napkins was conspicuous 57 hours after the birth of the child." If the disease resulted from some sort of congenital "'freak' of metabolism," could it also, just possibly, be hereditary? Well, maybe, but, "although brothers and sisters share the peculiarity, there is, as yet, no known instance of its transmission from one generation to another."[5]

Garrod did, however, discover one astonishing piece of information, which he reported in 1901. In several families with alkaptonuric children the parents were first cousins.[6] This was the information Garrod was trying to puzzle out, the clue he was trying to follow, when William Bateson came into his life late in the winter of 1901.

Just a few years younger than Garrod, Bateson was also the son of a distinguished father. William Henry Bateson had been a classics scholar and the master of St. John's College, Cambridge. Instead of Greek and Latin literature, however, the younger Bateson decided to devote his life to the study of biology, most particularly zoology and morphology.

Like Garrod's, Bateson's life was affected by a revolutionary book: not Bernard's *Introduction to the Study of Experimental Medicine,* but Darwin's *On the Origin of Species.* Bateson was a member of the second generation of Darwinians. Like many of his colleagues, he knew that there were several holes in Darwin's original argument. Also, like many of his colleagues, he hoped to be the person who would become famous for successfully plugging at least one of them.

Darwin had argued that every population of animals and plants exhibits minute variations; some finches, for example, have shorter beaks than others. Some of these variations, Darwin had said, could be shown to be more adaptive than others. Short beaks would allow finches to grasp more kinds of seeds, which meant that they would be more likely to survive the struggles of life and, therefore, more likely to leave offspring. *If* those adaptive variations were hereditable and inherited, then, over the course of many generations, a new species, such as short-beaked finches, would arise. But that was a very big *if;* neither Darwin nor anyone else knew for sure whether variations were inheritable or, if they were, which would be inherited. That was the puzzle that Bateson and many others were trying to solve as the nineteenth century drew to a close.

In order to solve it, Bateson became an animal and plant breeder, the better to study patterns of both variation and inheritance. "He ransacked museums, libraries, and private collections," his wife reported; "he attended every sort of agricultural show, mixing freely with gardeners, shepherds and drovers, learning all they had to teach him." Not long after he married, he bought a home with a large garden on the outskirts of Cambridge. His wife worked with him, planting, collecting, and recording: "We had poultry-pens well stocked; we had row upon row of peas, poppies, lychnis; the garden was full of big and little experiments—some, tentative trails of subjects; some, serious undertakings . . . from the merest menial drudgery to high flights of scientific speculation, hand and brain were hard at work."[7]

For about a decade nothing much came of these painstaking breeding studies. Then, in 1900, Bateson read a paper that had been published thirty-five years earlier by an obscure Czechoslovakian priest and plant breeder named Gregor Mendel.[8] Like Bateson, Mendel had been rigorously breeding plants: ornamental sweet peas. Unlike Bateson, however, Mendel had focused his attention on sweet-pea variations that were dichotomous, varieties that had either/or traits. Sweet-pea varieties are *either* tall *or* short; some have *either* green seeds *or* yellow; and some have *either* wrinkled seeds *or* smooth. Mendel had crossbred a large number of plants, keeping very careful track of the results of his crosses, the second generation. He then grew an even larger number of third-generation plants from the seeds produced by the second.

In the end, because he had recorded thousands of observations, because he knew the laws of probability and statistics (he had studied mathematical physics for a while as a university student), and because he had chosen his experimental variables with such care, Mendel was able to discern certain patterns in his data, patterns that had eluded Bateson and hundreds of other naturalists.

Although his path to the discovery was both painstaking and arduous, the patterns Mendel discovered can be expressed quite simply. First, in each pair of dichotomous traits, one always dominated over the other: when short was crossed with tall, all the resulting plants were tall; and when green-seed was crossed with yellow-seed, the seeds of all the resulting plants were yellow. Second, and equally important: when variations passed from one generation to the next, each member of the pair of traits acted as if it were an independent "factor" (the English translation of the German word Mendel used), passing to the next generation as if it were independent of its twin. This is why, when the yellow-seeded plants of the second generation self-fertilized, one-quarter of the resulting third-generation progeny reverted to the original seed color, green. (Today we would refer to this as a 1:2:1 ratio; one-quarter homozygous for the dominant trait, yellow seed; one-half hybrid, or heterozygous, but showing the dominant trait, yellow seed; and one-quarter homozygous for the recessive trait, green seed.)

Bateson's "delight and pleasure," his wife recalled, "on his first introduction to Mendel's work were greater than I can describe; as when with a very long line to hoe, one suddenly finds a great part of it already done by someone else and one is unexpectedly free to get on with other jobs."[9] Bateson quickly translated Mendel's paper into English and rushed it into publication. He began reorganizing his own experiments to see if he could find Mendelian ratios in other plants and in his chickens. And, as was his wont, he avidly read journals and talked to everyone he knew, looking for other examples of Mendelian inheritance.

This must be why he was introduced to Archibald Garrod—who

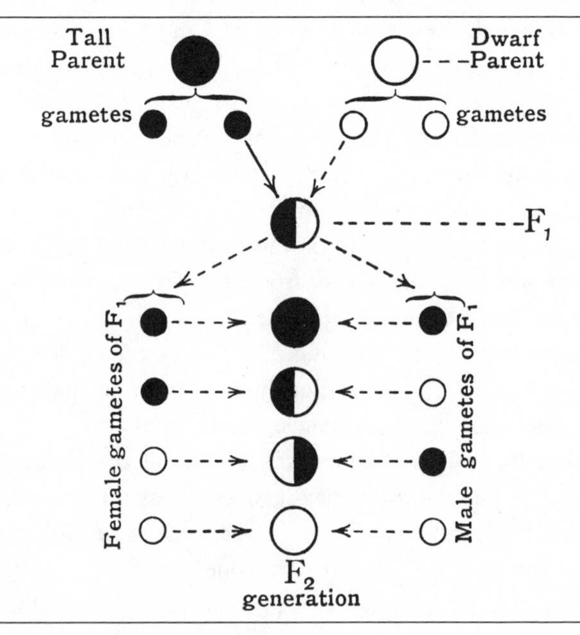

An early depiction of Mendel's Laws.

From R. C. Punnett, *Mendelism* (1922), 19.

was himself struggling to make sense out of the fact that people with perfectly normal urine had children (often more than one) whose urine was black.

Both men realized right away what Garrod's findings meant: alkaptonuria was a Mendelian recessive. Bateson understood that with Garrod's data on alkaptonurics he could extend Mendel's laws to human beings. Garrod understood that he had found a new type of etiology, disease by inheritance. By 1902 both men already suspected that albinism, hemophilia, and cystonuria (crystalline sediment in urine, which sometimes forms stones that remain in the bladder) were heritable conditions; by 1909 Garrod had added pentosuria (high levels of the pentose sugars in urine) to the list of illnesses acquired through inheritance.

Neither Bateson nor Garrod had much of an interest in the eugenic implications of what we would now call genetic disease. Bateson did once give a lecture to the Eugenics Education Society, but he began by admitting that he had never had much interest in its activities, "nor even to become a member."[10] Garrod's frame of mind was even further removed from eugenics and policy matters, as indicated by the subtitle of his 1902 paper, "The incidence of alkaptonuria: A study in chemical individuality," and the main title of his 1909 book, *Inborn Errors of Metabolism*. The only applications of his findings that mattered to Garrod were those which shed light on the nature of human metabolism, what would today be called basic medical science. Garrod believed that each person has a unique biochemical or metabolic personality, a personality that is at least partially inborn, or determined before birth: "Just as no two individuals of a species are absolutely identical in bodily structure neither are their chemical processes carried out on exactly the same lines."[11] Beyond this conclusion, Garrod was unwilling to speculate. Such a conclusion would have been anathema to the sort of eugenicist who sought diagnostic criteria by which whole groups of afflicted individuals could be deemed to be either "fit" or "unfit."

Thus the founding fathers of medical genetics were a physician interested in biochemical research and a geneticist interested in evolutionary theory—neither of whom had any particular enthusiasm for eugenics as it was being practiced in their day.

Medical Genetics

Some scientific fields take off very quickly (stem cell research is a recent example) while others begin slowly, dawdling for a time after the starting gun is fired. Medical genetics falls into the latter category. Born in the early years of the twentieth century, it did not mature until after World War II—a long time in comparison to other then-young sciences such as atomic physics or bacteriology. The impressive conclusions that Bateson and Garrod had reached did not bear immediate fruit. When medical geneticists today look back on the history of their field, they find this lag astounding. How could physicians and biologists have ignored for so long the fertile frontier that Bateson and Garrod had begun to explore?

The convoluted and potent politics of eugenics provides only part of the answer to this question. Other parts arise from the practical difficulties of studying human genetics and from the sociology of scientific disciplines. Young scientific fields require scientific entrepreneurs, people filled with energy and commitment, to get them up and running. Neither Bateson nor Garrod cared to act as such a founding father for medical genetics. Bateson had taken up genetics because he wanted to understand the relationship between hereditary mechanisms and natural selection: because he wanted, in short, to solve the riddles that had stumped Darwin. He was not a physician and he was not much interested in diseases.

Garrod too might have created medical genetics had he cared to, but for a variety of reasons he did not. Garrod's discoveries were made very early in his career. As he aged, his duties as a physician became more burdensome and he found less time for research; his list of publications contains very few research papers after 1910.

By the time he became Regius Professor of Medicine at Oxford in 1920, a post that relieved him of clinical responsibilities, personal tragedy had dampened his spirit. Garrod and his wife had four children, three boys and a girl. All three boys died in uniform during World War I. After the war, as Alexander Bearn, the only biographer who has had access to Garrod's private papers, put it, "Garrod's enthusiasm was forever quenched, and he became a reserved and private person."[12]

In any event, Garrod's fundamental interest in the "inborn errors of metabolism" was metabolic, not genetic. His hopes for the future of medicine rested on explorations of the metabolic uniqueness of each patient, not the familial pattern of disease incidence. He advocated biochemical testing, not the construction of extensive pedigrees. Toward the end of his life Garrod was invited to a meeting to discuss creating a British Council for Research in Human Genetics, but he declined to attend. "Garrod was not swept off his feet by genetics," Bearn concludes. "Clinical chemistry and the principle of biochemical individuality had continued to advance understanding of disease, but *genetics seemed to hold little clinical promise.*"[13]

Classical Genetics

Another reason why medical genetics failed to coalesce as a discipline in the early decades of the twentieth century was, ironically, that genetics itself was flourishing, drawing the attention of hundreds if not thousands of young investigators, men and a few women with a well-developed sense of where new scientific discoveries, and hence reputations, were likely to be made. Three research techniques made this growth in genetics possible.

The first was controlled studies of breeding. Mendel had shown how much could be learned from carefully controlled breeding studies, and others were quick to follow in his footsteps. Agricultural scientists were the initial enthusiasts. Many already had facilities for large-scale breeding, and by 1910 they had demonstrated that a host of important traits in a host of important organisms—wheat, corn, and chickens, for example—were Mendelian. Fortunately (or, some would say, unfortunately), controlled breeding studies could not be done on human beings, either then or now.

The second technique was cytological analysis. Techniques for staining and observing the contents of cell nuclei had been yielding up many secrets since the 1880s, when several German cytologists had unraveled the existence and behavior of stringlike bodies inside the nucleus. They called these bodies "chromosomes" because of the ease with which they could be dyed. These scientists also discovered "mitosis," the strange but wondrously orderly minuet that chromosomes perform when cells divide: first duplicating themselves, then lining up in the center of the cell, then moving far apart in two different ranks, one copy of each chromosome in each rank. Mitosis ensured that when the nucleus and the cell divided, separating the ranks, each resulting daughter cell would have the same number of chromosomes as the mother cell from which it had originated.

Before the turn of the twentieth century, cytologists had discovered that mitosis is almost ubiquitous. The cells of birds do it and so do the cells of bees and humans—which meant that human cells also undergo mitosis. In fact, *all* cells do it, except one type of cell on one special occasion. Late-nineteenth-century cytologists discovered that the reproductive cells of multicellular organisms do not undergo mitosis in the very last division that brings them to reproductive maturity. In that cellular division, the chromosomes divide *but they do not duplicate*. Cytologists called this "reduction division," or "meiosis," because the resulting daughter cells ended up with only half the number of chromosomes as the mother cell. Nature had turned out to be extraordinarily clever. Because of meiosis, the new organisms that resulted from the merger (at fertilization) of two reproductive cells would then have a full complement of chromosomes: not half a complement and not twice a complement, but just the right, species-specific number.

The species number for human beings turned out to be very high; so many chromosomes were packed into so small a space that it was impossible—until the 1950s—even to determine precisely how many there were. As a consequence, conclusive cytological studies could not be done on humans during the first half of the twentieth century.[14]

The third technique, and the most crucial for the growth of genetics, was a combination of the first two. After Mendel's work was rediscovered in 1900, it did not take very long for someone to figure out that chromosomes behave in the cell the same way Mendel's factors behave in the whole organism. Walter B. Sutton, a graduate student in cytology at Columbia University, published a paper in 1903 in which he argued that since both the chromosomes and the factors independently assorted, then the factors must be, in some way, carried on the chromosomes. Mendel had said that for each trait there must be two factors (a dominant and a recessive—Sutton called each member of the pair an "allelomorph") that sort themselves out in reproduction, with one factor going to each offspring. The cytologists had demonstrated that the chromosomes come in pairs and that, in meiosis, one member of each pair goes to each offspring cell, where it combines with chromosomes from the other parent.

Shouldn't it follow, then, that the factors, or allelomorphs, or whatever determined the traits, were *on* the chromosomes? It should also mean, in Sutton's words, that since "the chromosomes permanently retain their individuality, it follows that all the allelomorphs represented by any one chromosome must be inherited together."[15]

Sutton's hypothesis suggested that cytological analysis (the study of the chromosomes) could be combined with controlled breeding studies to yield enormous amounts of information about heredity. Indeed, within a few years a whole new vocabulary had been developed based on this combination of techniques, signifying that an entirely new scientific discipline had been born. Bateson first used the term "genetics" to describe controlled breeding studies in 1905. A few years later a Danish botanist, Wilhelm Johanssen, suggested that the determinants of traits (Mendel's factors) should be called "genes," although he admitted having no idea what the genes actually were.

"Allelomorph," Sutton's word for the different forms of traits carried on any single pair of chromosomes, was eventually shortened to the word used today: "allele." Subsequently, following a suggestion made by Johanssen in 1911, the whole set of genes in any given species (what the cytologists could imagine being carried on the chromosomes) was called the "genotype," and the whole set of traits that they determined (what the breeder studied when artificially mating organisms) was called the "phenotype."

As Sutton's hypothesis was coming to be accepted, many research scientists began looking for an organism on which it would be easy to perform controlled breeding experiments with cytological follow-up. *Drosophila melanogaster,* the common fruit fly, turned out to be the successful candidate. *Drosophila* can live in laboratory bottles and requires nothing but mashed bananas for food. These flies reproduce very rapidly and have large numbers of offspring in each cycle. The species has only four chromosomes, which are quite large. Under the microscope, scientists can see the fly's chromosomes in the cells of its transparent salivary gland, without even having to perform a dissection.

In 1910 Thomas Hunt Morgan, then a professor at Columbia University, published a paper in which he used breeding and cytological studies of *Drosophila* to provide convincing proof of Sutton's hypothesis. The experiment demonstrated that one particular *Drosophila* trait (white eye color) was carried on one particular *Drosophila* chromosome, the one that determined sex.[16] This experiment was the paradigmatic case for what came to be called classical genetics, the set of notions articulated in a book Morgan wrote with several of his students and published in 1915, *The Mechanism of Mendelian Inheritance.* According to the paradigm, the genes are laid out on the chromosomes like beads on a string. The genotype, which determines the phenotype, not the other way around, is located in the cell nucleus. Genes are lined up on each stringlike chromosome in a linear fashion. Most of the time, genes and traits exist in two versions, the alleles; one usually dominates over the other in the phenotype. Chromosomes are the physical bodies that are passed from one generation to another, creating the patterns of inheritance; the behavior of the chromosomes explains what the German cytologist August Weismann called "the continuity of the germplasm." Genes can occasionally change—or mutate. Mutations seem to be caused by environmental factors, but this is the only way the environment can influence the passage of traits between one generation and the next. Traits acquired during the life of the organism cannot be inherited, as Lamarck once believed.

Armed with this paradigm, the first generation of geneticists made spectacular progress in understanding the mechanisms of heredity. By 1920 more specific *Drosophila* traits had been discovered on specific *Drosophila* chromosomes. Observations of crossing over—the fact that when chromosomes are packed together in the nucleus they sometimes break and recombine so that part of one chromosome gets attached to part of another—had also been used to estimate the distance between some genes on an individual chromosome. (Because of crossing over, some traits that are usually linked to each other become uncoupled. If you mate two flies that have coupled traits and then compare the number of their offspring that have uncoupled traits with the number that continue the coupling, you end up with an estimate of how far apart the two genes are on the chromosome.) Some genes had actually been mapped; distance estimates based on crossing over had been correlated with the physical features of chromosomes seen under high-powered microscopes.

In science, as in other aspects of social life, success breeds more success—and this is part of the reason why the growth of classical genetics hindered the progress of medical genetics. For thirty years, roughly from the publication of *The Mechanism of Mendelian Inheritance* in 1915 until the end of World War II, classical genetics flourished. Research laboratories were established all over the world. Young scientists flocked to the field. Results poured out

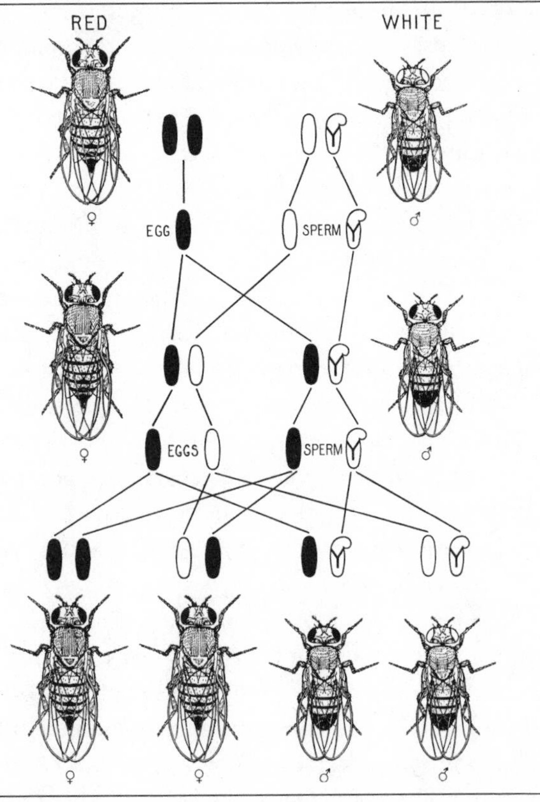

Depiction of Morgan's paradigmatic red eye–white eye cross, which demonstrates that in *Drosophila* the gene for eye color must be located on the Y chromosome.

From R. C. Punnett, *Mendelism* (1922), 105.

of laboratories. New journals were established. Successful applications—especially in agriculture—multiplied. Nobel Prizes were awarded.

Human Genetics

Under the banner of eugenics, human genetics had started out with great enthusiasm and very high hopes in the early decades of the century. By the 1920s, however, many of those hopes had been dashed, and a brain drain was starting to occur. The best and the brightest biologists were flocking into *Drosophila* genetics, leaving human genetics as something of an orphan discipline, with fewer and fewer active researchers and proponents. Human genetics did not, indeed could not, adopt the productive combination of controlled breeding studies and cytological analysis that had caused classical genetics to flourish; it could not produce verifiable results. Young biologists who wanted to make their mark, to make concrete discoveries that would lead to successful careers in a new science, preferred to work with *Drosophila.*

Many of the geneticists who remained committed to eugenics were biostatisticians, mathematicians by training and/or inclination, not biologists or breeders. After 1918 the biostatisticians began shaping human genetics in a fashion that dismayed some physicians and disgusted others. This is another part of the reason why medical genetics, a partnership of medicine and genetics, developed slowly.

In the waning decades of the nineteenth century, the study of human genetics depended on two research techniques: the first was the construction of family pedigrees, and the second was large-scale studies of the distribution of traits in human populations. Francis Galton, the founder of the eugenics movement, was one of several investigators who pioneered the use of both techniques. Galton published and distributed books of pedigree forms, instructing physicians to fill them out for their patients and mail them back to him. He also established what he called an "anthropometric" laboratory at the South Kensington Science Museum in London. Families visiting the museum could pop in to his laboratory to be measured, providing him with thousands of data points comparing, for example, the height or chest breadth of parents and their children.

Unlike Bateson and Garrod, Galton had the energy, the commitment, and the money to create a discipline based on his research techniques. He established a laboratory and a professorship at University College, London—both of which were utilized by his most devoted disciple, Karl Pearson, to create the discipline that is now called biostatistics. Pearson had been trained as a mathematician and was teaching mathematics at University College when he first became entranced by Galton's statistical studies of human heredity. Pearson immediately became a eugenicist in all of Galton's original senses of the term: he wanted to develop Galton's statistical method of correlation analysis so as to study, rigorously, the variability of human populations *and* so as to develop scientifically grounded social policies that would guide human evolution in what he thought to be positive directions.

Pearson was as energetic as Galton. He was able to raise substantial amounts of money from eugenically inclined philanthropists, money he could add to the already munificent Galton endowment. By 1920 the Galton Laboratory had been able to hire half a dozen researchers and to collect and analyze huge amounts of data about, for example, the incidence of tuberculosis, diabetes, and alcoholism in the British population. The laboratory was also publishing its own journal, *Biometrika* (still the leading journal in the field today), as well as its own series of eugenic pamphlets and scientific monographs.

Early in his career in genetics, Pearson had a very unpleasant argument with William Bateson, one with important historical ramifications. At first the dispute was about evolutionary theory, but it soon expanded to the validity of research methods, and that is when it became particularly hostile. Bateson, as we have seen, was a passionate advocate for Mendel's laws; Pearson was equally passionate about a somewhat different theory of heredity, Francis Galton's law of ancestral inheritance. Each man thought that if one law was correct the other had to be erroneous. Bateson, like Mendel, believed the determinants of traits were discontinuous (meaning that they existed in alternate or separate states that were very different from each other, like brown and blue eyes or short and tall plants); Galton and Pearson thought they were continuous (meaning that the determinants blended with each other to produce small variations in traits, like human height or finch beak length). Bateson thought evolution had to proceed in big steps, through discontinuous variations (mutations) that were heritable. Pearson thought evolution had to proceed in tiny steps, through natural selection of minute variations.

That was the initial, theoretical, argument. It might not have amounted to very much (scientific fields have been known to survive such disagreements; indeed sometimes they emerge the better for them) had it not turned into a dispute about whose scientific results could be trusted, about who was the better geneticist, possessed of the better research method. At that point the argument turned both personally ugly and scientifically damaging.

The fundamental problem arose from the fact that Mendelian laws are probabilistic. They are based on the assumption that assortment, or the segregation of alleles, is independent, or essentially random; chance, and only chance, determines which daughter cell gets which allele, the same way chance determines which face of a coin faces up after a toss. If you throw a nickel into the air two thousand times you are likely to come up with something close to one thousand heads and one thousand tails, but if you toss it only ten times, five of each is very unlikely. The probability of getting a 1:1 ratio is calculable; it depends on the number of throws, and it gets higher the more throws there are. Because Mendel guessed that inheritance was probabilistic, he knew that he would have to examine thousands of sweet-pea plants and thousands upon thousands of seeds in order to see any pattern at all. If he had only looked at a few parent plants and a few offspring plants, he probably would not have been able to see any patterns or generate any laws. For the same reason, Morgan knew that he had to look at hundreds of fruit flies in order to determine whether, for example, red eye or white eye was the dominant trait. Indeed, the probabilistic nature of Mendel's laws is precisely the reason that geneticists of Morgan's generation sought experimental organisms, like *Drosophila,* that bred frequently and had many offspring in each generation.

The first Mendelian eugenicists—human as opposed to experimental geneticists—did not understand the depths of this problem. Fired with enthusiasm, they went about collecting pedigrees for any number of human traits that they hoped would turn out to be inherited and then tried, using Mendel's paradigm, to figure out which of the traits were dominant and which were recessive. Their efforts were quickly rewarded with regard to sex-linked recessive traits (like hemophilia and colorblindness) because they could see, even in a small number of pedigrees, that men were afflicted and that women were not afflicted but acted as carriers. Unfortunately, in almost all other cases—diseases, conditions, and traits that were not sex-linked—the results they obtained, based as they were on only a small number of pedigrees, were often inconsistent: in one family pedigree brown would appear to be a dominant eye color, and in another blue would look dominant; one person with diabetes could trace the disease in her family for six generations, while another had no relatives with the disease.

People who understood probability and statistics understood why these inconsistencies had to arise from the study of a small number of pedigrees, but most biologists of the day had no training in higher mathematics, let alone in probability and statistics. Karl

Pearson was, however, a mathematician, as were several of his colleagues in the Galton Laboratory. They were also all eugenicists, but—and this is the crucial point—they were not Mendelians. As devoted as they were to eugenics, they were equally devoted to the scientific discipline they were in the throes of developing: biostatistics, the statistical analysis of large quantities of populational data.

Thus if a Mendelian eugenicist, often a colleague or friend of Bateson's, published a paper (based on pedigrees) demonstrating that eye color or feeblemindedness or cancer or diabetes is inherited in a dominant/recessive pattern, a few months later a Galtonian would publish a paper saying that there were not enough cases to be certain, or that the pedigrees contained vastly too much anomalous or vague information, or that the Mendelian author had to have fudged his (or, occasionally, her) data. To make matters worse, Galtonian papers were written in the mathematical language of probability and statistics and often contained jibes about the scientific flabbiness of qualitative (for which the audience was supposed to read "Mendelian") as opposed to quantitative (read "biostatistical") work.

"Scientific flabbiness" is a fighting phrase—and the fight was soon engaged. Pitched and public battles ensued, nasty words were spoken at scientific meetings, papers were turned down for publication in respected journals, and candidates for jobs were grilled about their predilections. Unaffiliated scientists found it very difficult to mediate or to collaborate with both parties.

There were two important consequences of this acrimony. First, in the early decades of the twentieth century very few conclusions drawn from pedigree analysis of human traits went unchallenged. Most of the challenging biostatisticians were as devoted to the eugenic cause as their Mendelian opponents, but they had to admit that the evidence—particularly for the inheritance of behavioral traits such as feeblemindedness and alcoholism—was exceedingly weak. Second, in their frustration with the difficulties of doing human genetics, the second generation of biostatisticians, those who came to maturity after World War I, retreated from the field of human genetics almost entirely.

The biostatisticians had cut their statistical teeth demonstrating that Mendelian studies of pedigrees were faulty, but eventually they also had to acknowledge that the research technique favored by Pearson, the collection and analysis of large quantities of populational data, presented serious methodological difficulties of its own. With painstaking and tedious effort you could discover the variability in a trait in a given population, for example, how many people in a particular neighborhood, served by a particular clinic or hospital, had been diagnosed with TB (or cancer or diabetes or alcoholism). Then you would have to make even more strenuous efforts to divide that population into generations, so as to study the inheritance of the variation. If you did come up with this data, you could use statistical techniques to determine the strength of the correlation between the incidence in the two generations. However, and this was the rub, when you were all done, you still could not know whether the correlation you had just calculated was caused by the environment or by the genes; by nurture or by nature. Since the individuals in those two generations in the same neighborhood had, presumably, shared their environments, how could you know, for example, whether the high correlation between alcoholism in the parents and alcoholism in the children was the result of a gene for alcoholism, or of the fact that young people learn to drink from their parents? The biostatisticians realized that such crucial questions could only be answered by studies of identical twins reared apart—and where, they wondered, were they going to find a sufficient number of such sets of twins to study?

As a consequence of these difficulties, human genetics began to seem far too recalcitrant to the younger generation of biostatisticians. Their enthusiasm for eugenics did not falter, but their conviction that they could learn enough about human heredity—enough to propose successful genetic solutions to real social problems anytime soon—began to dissipate. In their frustration they turned, again, to problems in the Darwinian theory of evolution, either pursuing purely mathematical investigations or collecting data on organisms with which it was possible to do controlled experiments.

In 1918 the mathematically keen biologist R. A. Fisher had demonstrated that there might be a resolution to the original Bateson-Pearson argument about whether discontinuous (Mendelian) or continuous (Galtonian) variations served as the basis for evolution. Fisher's analysis suggested that *both* kinds of variation were Mendelian, that both were important for evolution, that continuous variation probably resulted from the interaction of a large number of alleles, and that both continuous and discontinuous variation could be successfully studied in large populations using the implications of the Hardy-Weinberg law ($p^2 + 2pq + q^2 = 1$, where p is the frequency of the dominant allele and q is the frequency of the recessive).[17]

Fisher's insight turned out to be extremely fruitful, so much so that by the end of the 1930s it had produced what has come to be called the neo-Darwinian synthesis, a set of ideas that—among other things—plugged up the old, plaguing holes in Darwinian theory, the ones that had galvanized Bateson's interest even before the turn of the century. Fisher's insight also led to the creation of a series of rigorous techniques (of what came to be called population genetics) that enabled biostatisticians to estimate the frequencies of individual genes in different populations. To put the matter another way, biostatistical techniques gave human geneticists a method for understanding the genotype when all they could see (and count) were the phenotypes in a population.

This had been the fundamental problem faced by human geneticists: lacking the ability to do controlled breeding experiments with cytological follow-up, they needed some other research technique by which to connect phenotypes with genotypes, something that neither pedigree studies nor populational data could do very well. By the mid-1930s, armed with neo-Darwinian techniques, biostatisticians felt ready to attack problems in human genetics again. But the genetics they developed, the genetics of populations, was a very different science from the genetics with which the original, Mendelian eugenicists had begun. And, crucially for the history of medical genetics, it was a science that most physicians regarded as either incomprehensible or irrelevant—or both.

In order to study human genotypes, populational human geneticists needed phenotypes that were both easy to define and easy to sample; they found what they were looking for in the human blood types. The various blood types (A, AB, B, and O), which had been discovered about twenty years earlier, had been shown to be inherited in a more or less straightforward Mendelian fashion. Blood samples were relatively easy to obtain, and blood types were relatively easy to define; all that was required was a test for the presence of a particular antigen in each sample. Analysis of the blood samples of large numbers of people revealed that different populations—Germans, Scots, white Americans, black Americans—contained different proportions of the blood types.

Biostatisticians and eugenicists made much of this discovery; the biostatisticians because it proved that neo-Darwinism could be extended to human populations; the eugenicists because they hoped (though the hope would not be realized) that they now had a marker that would enable them to separate the various races of man. Unfortunately, the discovery was incomprehensible to most physicians, couched, as it was, in the abstract languages of population genetics, algebra, and evolutionary theory. Archibald Garrod surely spoke for most of his medical colleagues when he confessed, in a letter written in 1933: "I find myself quite out of my depth, in the new Mendelism of Hogben and Haldane [two prominent English biostatisticians]".[18]

Physicians' Attitudes toward Human Genetics

Lack of comprehension was not the only reason physicians ignored human genetics. They also thought it to be medically useless. Yes, it was important to know which blood type a person had, especially if the person needed a transfusion, but what difference did it make, either to treatment or to cure, to know what *proportion* of your patients were A, AB, B, or O—and whether that proportion was likely to change in succeeding generations? By becoming the study of gene frequencies, populational human genetics had become medically irrelevant.

On top of this, a fairly large group of elite physicians regarded eugenics—the political ideology to which human genetics, in all its guises, was still wedded—as anathema. Many of these were progressive physicians, who advocated for and avidly pursued scientifically grounded medical research. They were also prominent in the medical community. They were people who had clinical appointments in teaching hospitals and who determined the curriculum in medical schools.

In the first half of the twentieth century, medicine was experiencing a "golden age"; physicians were developing and applying effective diagnoses, therapies, and even cures for some devastating diseases. The two sciences that lay at the foundation of this medical revolution were bacteriology and what was then called clinical chemistry (biochemistry is the name used today). Bacteriology yielded the germ theory of disease—and the acknowledgment that human scourges such as tuberculosis, syphilis, childbed fever, and pneumonia were caused by minute organisms that could travel, unseen, from one place to another, from one person to another. Biochemistry yielded the chemical theory of disease—and the acknowledgment that other human scourges, including diabetes, scurvy, and rickets, were caused by imbalances in certain identifiable, essential chemicals or by errors in metabolic and hormonal pathways. Armed with new scientific diagnostic tests (for iron in the blood, sugar in the urine, spirochetes in the feces, or amoebas in the water), physicians felt they were finally able to make sense out of some diseases and to distinguish diseases, not by their symptoms, but by their causes. Armed with new scientific therapies (such as salversan for syphilis, antitoxin for diphtheria, insulin for diabetes, iron for anemia, vitamin D for rickets), they were excited to be able to offer their patients, and themselves, some sense of hope. Many early-twentieth-century physicians experienced the hubris of the successful; they felt personally powerful, medically effective, and socially respected.

Not surprisingly, many of them never warmed to one of the eugenicists' fundamental messages: that scientific medicine was having negative effects on the human race. Eugenicists—no matter what their methodological persuasion—made no secret of their disdain for the accomplishments of their contemporaries who were physicians. Modern medicine harms the human race, they argued, because it sustains the unfit, allowing them to live and reproduce, when it would be better to let them die out. To use a phrase that started to become popular among geneticists in the 1930s, new medical practices were increasing the "genetic load," the incidence of "bad genes" in human populations.

"Everything which tends to check the multiplication of the unfit," Pearson asserted in 1912, "and to emphasize the fertility of the physically and mentally healthy will . . . aid Nature's method of reducing the phthisical [tuberculosis] death rate . . . £1,500,000 spent in encouraging healthy parentage would do more than the establishment of a sanatorium in every township." In a lecture given to a conference of public health physicians in the same year, Pearson lambasted his audience, not only for failing to collect good data on public health, but also for wasting their time on sanitary reform. "The health of the parents," he insisted, "is far more important than the question of back-to-back houses, the apartment tenements, the employment of mothers, or the practise of breast feeding." Pearson had earlier enraged the medical advocates of prohibition, both in Britain and in the United States, when he publicly argued that temperance campaigns would never cure alcoholism because hereditary degeneracy, not alcohol, was the cause of alcoholism. Several years later, going right for the medical jugular, Charles Davenport, director of the Eugenics Record Office in New York, told an audience of medical students that many of the diagnostic tests about which the faculty were so pleased to be teaching them were, actually and profoundly, faulty—because they failed to separate genetic diseases from other types of illness: "Just as a pathology of 60 years ago looks crude today, so . . . that of today will look no less crude 60 years hence . . . [when] the properly trained physician will not regard the matter of [family] history taking as a frill or a useless concession to highbrow insistence."[19]

Small wonder, then, that in the first half of the twentieth century many physicians found eugenicists more than a little irritating. Even some physicians who had been swept up by eugenic enthusiasm between 1910 and 1920 lost the faith within a few years; so little progress had been made in medical applications of genetics while enormous progress had been made in the applications of bacteriology and biochemistry. Being able to make judgments about disputed paternity (the only application, then, of the genetics of the blood groups) was insignificant when compared with being able to immunize children against diphtheria or using antisepsis and anesthesia to ensure successful surgical outcomes. Similarly, being able to tell prospective parents that they should not have children because the wife's brother was a hemophiliac paled in comparison with being able to tell prospective parents that their babies need no longer die from diarrhea. Only a few human diseases (hemophilia) and a few human conditions (colorblindness, alkaptonuria, albinism) had been clearly defined as genetic. Eugenics ideologues had managed to convince a fair number of politicians that their science had produced usable results, but by 1940 most physicians, like most geneticists, knew better.

The men and women who struggled to create the field of medical genetics in the interwar years knew full well that physicians were resistant. In 1927 Lewellys F. Barker of the Johns Hopkins School of Medicine noted "the apparent apathy of medical men with regard to the problems of inheritance."[20] A decade later J. B. S. Haldane, the British biostatistician, remarked: "Doctors are not in general taught human genetics. A medical student who has attended three lectures on the entire subject of genetics is unusually well informed." In the early 1930s Laurence Snyder, a biologist, was appointed as one of the first professors of medical genetics in the United States, at Ohio State University, but he had difficulty convincing the medical faculty to take his work seriously: "I was asked publicly to explain the gene for a stomach, and to give an opinion on whether the gene for the heart was dominant or recessive," he later recounted. He eventually gave up research and proselytizing for the field and became a dean.[21]

Medical genetics did not flourish, did not begin to build upon Garrod's original insight, until a partnership formed between physicians and geneticists. The obstacles to that partnership did not begin to disappear until after World War II, that is, until the members of the generation of eugenicist geneticists were in retirement and their ideology was in disrepute. Before that time, both with regard to research and with regard to public health policy, eugenicists and physicians had very little in common. The eugenicists may have had some success in proselytizing among politicians, but among physicians they had made little headway despite initial enthusiasm. Some substantive research had been done about the inheritance of human phenotypes, but only a small portion of the phenotypes examined was of much interest to physicians, and very little of what the eugenicists even thought they knew could be applied to diagnosing, treating, or curing sick people. Equally as infuriating to physicians was the fact that the foundational ideology of eugenics demeaned and dismissed the progress that many physicians thought they were making and had worked very hard to achieve.

The Genealogical Fallacy and Eugenics

Many critics of what is today called medical genetics—the effort to develop, administer, and evaluate genetic tests with the intention of curing or preventing diseases and other disabling conditions—are fond of arguing that it was both morally and socially corrupted by its historical origins in eugenics. Such an argument commits what historians call "the genealogical fallacy": roughly akin to punishing the grandchildren for the beliefs of the grandparents, or, in this case, visiting the sins of the second cousins once removed upon their relatives, fifty years later, who had made a considerable effort to repudiate the original stain on the family name. Yes, these people are connected to one another, but their fundamental beliefs are not the connection.

While it is undoubtedly true that Galton was a eugenicist, it is not true that the biostatisticians who worked in the Galton Laboratory two or three decades after Galton's death believed it was a good idea to sterilize people who were mentally retarded; in point of fact, Galton did not believe that either. Charles Davenport was a Mendelian, who did indeed believe in the wisdom of sterilization, but that did not mean that most Mendelian geneticists of his generation were eugenicists. Hermann J. Muller was both a Mendelian geneticist and eugenicist, who won a Nobel Prize for his study of mutation rates, but that does not mean that other *Drosophila* geneticists, human geneticists, or even medical geneticists applauded his idea of creating a sperm bank for geniuses, or that they donated their sperm, or that they recommended the use of such sperm to their daughters.

We should not, in short, assume that because one member of a group had an idea, everyone else in the group agreed; even less should we assume that because one or more members of a group had an idea in one generation, their successors held the same idea in the next. If mapped out as its own pedigree chart, the first five or six decades of the history of genetics would look like a Darwinian phylogenetic tree, except that it would have three trunks: the Galton trunk would produce biostatistics, and then a branch of that would merge with the Mendel trunk to create population genetics. Human genetics would start out as a branch, a little lower down on the Mendel trunk, which would also merge with population genetics after it had grown a bit. The third trunk, medicine, would not have a branch connected to human genetics until all three trunks had gotten a little older. And if we colored the tree blood red wherever eugenics had dominated, only a small portion of the trunk would look as if it had been wounded; the rest would look healthy, because the infection had not spread.

The history of genetics is indeed connected to the history of eugenics, but not in a way that affects the social, medical, or moral project of medical genetics. To carry the analogy further, medical genetics does indeed have its roots in classical Mendelian genetics and in Galtonian biostatistics, but the fact that some (or, in the case of biostatistics, many) of the practitioners of those sciences were inspired by eugenic dreams does not mean that their dreams (or nightmares) were perpetuated along all of every branch. It is true, to specify just one example among many, that without the insights of both the Mendelians and the Galtonians, one of the first successes of medical genetics—determining the genetic etiology, the incidence, and the evolutionary advantage of the gene for sickle-cell anemia—would not have been possible by the mid-1950s. It is also true that, two generations earlier, most of the Galtonians and some of the Mendelians were eugenicists. But it does not follow—indeed it manifestly is not the case—that the people who made those discoveries were themselves either eugenicists or racists. Eugenic dreams may have inspired Karl Pearson to develop the Galton Laboratory and may have kept R. A. Fisher's nose to the biostatistical grindstone—and both men certainly left writings indicating that they regarded dark-skinned people as inferior to whites. But that does not mean that the men and women who used clinical tests for sickling hemoglobin and the mathematics of the Hardy-Weinberg formula to understand that the recessive gene had survived because it protected carriers from malaria undertook their research in order to figure out how to reduce the population of dark-skinned people.

Just short of four decades after Bateson and Garrod had started their collaboration, a young American Ph.D. candidate in classical genetics, James Neel, developed an interest in human genetics. After finishing his dissertation, Neel taught for two years and then spent a year on a research fellowship, continuing to study *Drosophila,* but that last year convinced him that *Drosophila* had supplied (for him, at least) all the interesting results it was ever going to yield. The next big frontier, he thought, was going to be human heredity, and to pursue it properly he would need a medical degree. "On December 4, 1941, just three days before the attack on Pearl Harbor," Neel recalled in his autobiography, "I wrote the University of Rochester, requesting admission as a second-year medical student." Neel was also convinced that being a physician would make him much more useful than being a geneticist if there was going to be a long war.

Neel knew that his decision to focus on human heredity would be regarded with suspicion by his fellow classical geneticists. "Many of the senior geneticists with whom I had discussed a move into human genetics . . . had almost perceptibly recoiled," he admitted in his autobiography. This was, he thought, "a reflex conditioned by the generally poor quality of 'research' on human heredity in the U.S. . . . and the opprobrium which the eugenics movement in the U.S. and Germany had rightfully earned." He decided to see for himself by spending the last month of his fellowship, before entering medical school, visiting the Eugenics Record Office in Cold Spring Harbor, New York.

> Only as I browsed through the files . . . did the magnitude of what had to be done before human genetics could be a respected discipline become brutally apparent . . . The concerns expressed by so many of my friends and advisors over the parlous state of human genetics . . . had been fully deserved. It was a real gamble to believe that I could bring the rigor of *Drosophila* genetics into this arena . . . [and] I wasn't going to have a lot of company in this gamble.[22]

That gamble paid off—and James Neel eventually had plenty of company. Because he was able to combine the rigor of a classical geneticist with the clinical acuity of a physician, Neel is now regarded as one of the founders of medical genetics, a field that has developed so extensively since that week he spent in Cold Spring Harbor that it now has thousands of practitioners who work in dozens of different clinical settings all over the world. But he and the other founders of medical genetics did something more than just studying human heredity rigorously; by developing reliable testing regimens for genetic conditions, they created a discipline (a good word here) which demonstrated that, in a therapeutic context, studies of human heredity could be purged of the political enthusiasms that had once led, not just to sloppy science, but also to violations of fundamental human rights.

CHAPTER 3

Pronatal Motives and Prenatal Diagnosis

Many medical professionals are known by the technologies they use: surgeons by their knives, dentists by their drills, pathologists by their microscopes. Indeed, some medical specialties owe their very existence to technologies. Where would anesthesiology be without anesthetics, or radiology without X-ray tubes and ultrasound machines? Genetic tests are the technologies that have helped the field of medical genetics grow. There were physicians, like James Neel at the beginning of his career, who practiced medical genetics before there were genetic tests, but their numbers were small and their influence barely detectable; the best they could hope to do for individual patients was to calculate the likelihood that a particular pregnancy might, or might not, produce an infant with a genetic disease. Once reliable tests had been developed, however, the presence or absence of a particular gene or a particular chromosomal abnormality in a particular infant or a particular fetus could be ascertained, which meant that parents and physicians could take effective action, either to prevent the birth of the fetus or remediate the disease in the infant. The availability of these tests resulted in a growth spurt for the clinical specialty: research programs expanded; funding levels increased; new professional associations and journals were created; residency programs were founded and medical students began applying for them; patients began demanding access to the new tests and physicians began offering them.

One of the first tests in widespread use was amniocentesis, a series of procedures for examining the tissues, cells, biochemical products, chromosomes, and, somewhat later, the DNA, not of an adult or a child, but of a fetus. Amniocentesis was the first of what are now several forms of prenatal diagnosis; chorionic villus sampling, another way of obtaining fetal cells, was developed somewhat later, as was ultrasound diagnosis, which assesses the condition of the fetus visually. All of these diagnostic tests can discern the presence of a genetic or congenital condition before birth, which means that they must be carried out on—and through—the body of a pregnant woman. As a result, they have wrought profound alterations in the experience of pregnancy for millions of women all over the world. Because the tests were developed and continue to be administered by physicians and other medical personnel, prenatal diagnosis has increased what has come to be called the "medicalization" of pregnancy, the conversion of a natural process into one that, if not exactly pathological, still must be mediated by highly trained professionals.

Prenatal diagnosis also raises a host of novel ethical issues including, but not limited to, the status of the fetus as a patient, the relationship between the fetus and its mother, and the relationship between the fetus and codes of law. The most common therapeutic option consequent on a prenatal diagnosis is termination of the pregnancy; as a result, these tests also raise difficult questions about the wisdom and the ethics of abortion.

Some opponents of genetic testing have argued that prenatal diagnosis is not the least bit different from old-fashioned eugenics; it enables elite experts to control the reproductive practices of ordinary people. Other opponents regard prenatal diagnosis as different from, but just as unethical as, Nazi eugenics, partly because the new genetic tests are both more accurate and more pervasive and partly because, in their view, abortion is a more serious crime than either sterilization or euthanasia.

Beginnings count, and so do motivations. Prenatal diagnosis is, like all medical tests, a technology, or rather a technological system: a linked series of procedures for examining the tissues, cells, biochemical products, chromosomes, and DNA of a fetus. Like all technological systems, prenatal diagnosis has a history, a history shaped by its beginnings and by the motivations of its founders. A great deal of light can be shed on the arguments of those who oppose prenatal diagnosis by careful examination of its history.

Prenatal Diagnosis as a Technological System

Amniocentesis, the original form of prenatal diagnosis, is a technological system in the same way that a can plus a can opener is, or a computer plus software and a printer: a successful outcome depends on the functioning of several linked components. In the case of amniocentesis the linked components are more numerous. The first is an amniotic tap, a process for getting amniotic fluid out of a pregnant woman's body. Then there are a set of procedures for looking carefully at the fetal cells that are floating in the fluid and at the fetally produced biochemicals that are dissolved in it. At the very beginning of its history, prenatal diagnosis could not examine the DNA in fetal chromosomes, but it could examine the chromosomes themselves, seeking out chromosomal abnormalities: breaks, absences, excesses. The technique for examining chromosomes is called karyotyping. Successful karyotyping requires standards or conventions for describing the number and shape of normal chromosomes: you cannot detect what is "abnormal" unless you are fairly sure that you know what "normal" is. Although we usually do not think of them this way, standards and conventions are technologies, which means that the standards and conventions for naming and describing chromosomes are part of the technological system of karyotyping and therefore part of the system—in fact a crucial part of the system, because, in the early days, they were the signs that allowed the experts to make a diagnosis.

Finally, a diagnostic system implies the existence of a possible therapy or a possible cure. The people who pioneered amniocentesis wanted to prevent the birth of fetuses afflicted with genetic or congenital disabilities and diseases. This means, of necessity, that there is one additional essential component of the amniocentesis system, a component that carries complex and freighted meanings in the space between its adjective and its noun: therapeutic abortion.

The Amniotic Tap

Historians of technology know that the chronological sequence in which technological systems develop is not always the same as the sequence in which consumers—in this case, patients—experience them. In the case of prenatal diagnosis, however, the first experiential step, the amniotic tap, was also the first chronological step. Recounting the story takes us back to the end of the nineteenth century and into the history, not of medical genetics, but of obstetrics.

The amniotic sac or amnion is one of the membranes that surround a fetus as it develops in the womb. Over the course of a pregnancy, fluid accumulates inside that sac; when a pregnant woman's "water breaks," it is actually the sac that has broken. The "water" that flows out is the fluid that has accumulated in the sac. Since the rupture of the amniotic sac is an entirely normal process, birth attendants were quite accustomed to breaking it themselves, if the necessity arose, which it frequently did. Some labors are unproductive; contractions can go on for hours without moving the fetus down the birth canal. In that circumstance, midwives and doctors used to rupture the sac, in hopes of speeding the delivery; indeed, a special instrument for the purpose was developed almost three hundred years ago, although any sharp instrument would do the job. Once labor had begun, inserting such an instrument through the vagina and cervix was a fairly easy task—and the patient did not experience pain (unless the instrument slipped) because the amnion has no nerve endings.

In the latter part of the nineteenth century physicians began puncturing the amniotic sac in efforts to alleviate a condition known as polyhydramnios, in which excessive fluid accumulates during the second half of a pregnancy. Modern studies suggest that about 2 percent of all pregnancies involve some version of the condition. Diabetic women have an elevated risk, although women who are not diabetic can experience it also. When severe, polyhydramnios is often fatal for the fetus; it can also be dangerous for the mother. Nineteenth-century physicians attempted to address this condition, tapping off some of the excess amniotic fluid by inserting a thin, hollow hypodermic needle through the pregnant woman's abdominal and uterine walls.[1]

By the 1930s the technique had acquired a name, "amniotic puncture." As insulin therapy began to help diabetics live to reach adulthood, many obstetricians became accustomed to doing amniotic punctures to extract excess fluid. The same procedure could also be used to inject something into the amnion. Distilled water was sometimes injected to alleviate oligohydramnios (too little amniotic fluid); salt water was sometimes injected to induce labor (in women who were overdue) or to terminate pregnancies in the second trimester. By 1950 amniotic puncture had become, if not a routine procedure, at least one that most obstetricians could undertake when necessary.[2]

Necessity came calling more frequently in the 1950s as a result of developments in the treatment of Rh disease. Birth attendants had known from time immemorial that some babies were born (often stillborn) with too much fluid in their tissues and their organ cavities, a condition known as hydrops. Other babies were born looking yellow, or jaundiced, and showed signs of brain damage in the first months of life: kernicterus. There were also some mothers who were unable, time after time, to give birth to live, healthy babies. In the early 1940s, as the result of some clever detective work by hematologists, obstetricians began to understand that these three phenomena were related by the presence or absence of a particular blood antigen, now called the RhD factor. If a mother was RhD negative and a father RhD positive, then with each pregnancy there was a 50 percent possibility that the baby would be RhD positive—that the mother and the baby would be RhD-incompatible. If mother and baby were RhD-incompatible and some of the baby's blood touched the mother's tissues during delivery (a common occurrence), then the mother's body would produce antibodies that would threaten all of her subsequent pregnancies in which the fetus was RhD positive. An immune reaction between her body and that of her fetus would damage the fetus's red blood cells, causing all kinds of developmental disabilities, including hydrops when the reaction was severe and kernicterus when it was less so.[3]

In white populations, approximately one in seven marriages is between a woman who is RhD negative and a man who is RhD positive. This means physicians in the developed countries of the world see a fair number of afflicted women and babies in the course of ordinary practice. By the 1950s they understood what was causing the problem and sought ways of coping with it. Early in the decade D. C. A. Bevis, an obstetrician living in Manchester, England, published a series of papers in which he first guessed and then demonstrated that in pregnancies at risk for Rh disease, the extent of damage to a fetus could be estimated by optical examination of amniotic fluid obtained by amniotic tap before birth.[4] In subsequent publications Bevis pointed out that physicians who obtained and used amniotic fluid in this way could estimate whether it was worth inducing labor early to mitigate the harm to the fetus and its mother. Alternatively, a physician armed with this information could prepare for a transfusion, a delicate and difficult but sometimes successful procedure when carried out on a newborn.

Because of what came to be known as the "Bevis test," one obstetrician reported, the mortality rate in his hospital for infants born to RhD-sensitized mothers had fallen from 22 percent in 1957–58 to 9 percent in 1962.[5] By 1960 several articles had been published

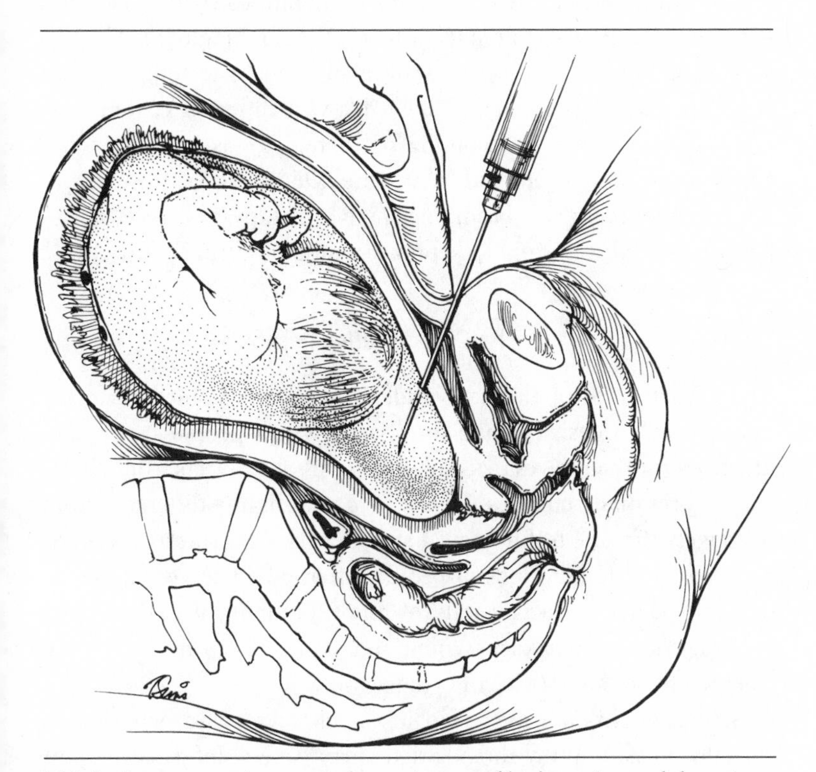

Third-trimester amniocentesis for treatment of hydramnios and detection of Rh disease.

From J. A. Pritchard and P. C. MacDonald, eds., *Williams Obstetrics*, 16th ed. (1980), 332. With the permission of The McGraw-Hill Companies.

providing evidence of the relative safety of the procedure and suggesting improvements in technique. The first of these articles, published in 1958, suggested changing the name from "amniotic tap" to "amniocentesis."[6] Thus the name we now give to a whole technological system was originally created as a name for only the first element in that system, the amniotic tap itself.

The two life-threatening conditions to which amniocentesis offered at least a partial solution, polyhydramnios and Rh disease, are managed by other means today: polyhydramnios by diuretics and Rh disease by postpartum injections of Rh immunoglobin. The history of technology is full of instances in which a technique developed for one use is applied to achieve a very different goal even as the first application is becoming obsolete. This is precisely what happened to amniocentesis. In the late 1960s, just as postpartum Rh immunoglobin was becoming the Rh therapy of choice, obstetricians started using amniotic taps for another purpose entirely.

The Cytogeneticists

The French anthropologist Claude Lévi-Strauss coined the term "bricolage" to designate the process by which people create new identities for themselves by combining symbolic objects from different cultures or subcultures. I use it here in a slightly different sense, to suggest that the technological system of amniocentesis was constructed by different kinds of people, often pursuing very different technological and social goals. Amniotic tapping was developed by obstetricians, but karyotyping and standardized nomenclature emerged from the practice of cytogenetics.

Cytogeneticists are interested in the character and behavior of chromosomes, the tiny threads that carry genetic information and are sometimes visible in the nuclei of dividing cells. Unlike obstetricians, cytogeneticists are "bench" scientists; they do their work behind the scenes in laboratories, not in clinics. The cytogeneticists who developed systematic observations of and standardized nomenclature for human chromosomes worked in laboratories in many different countries, in pursuit of knowledge about many different kinds of human diseases.

One of them was T. C. Hsu. Hsu was born in China in 1917 and arrived in the United States in 1948—just as the bloody civil war in his homeland was coming to an end—to begin graduate work in genetics at the University of Texas in Austin, a prominent center of *Drosophila* studies. When Hsu finished his degree, one of his professors helped him find a postdoctoral position at the University of Texas Medical Branch in Galveston.[7]

Hsu was hired because the medical laboratory needed someone to examine the behavior of chromosomes in cancerous tumors; in 1951 medical researchers were just beginning to wonder whether chromosomes could be responsible for the exploding growth rates of malignant tumors. Although Hsu had never examined a human chromosome, both his mentor and the head of the Medical Branch laboratory assumed that the techniques he had learned on *Drosophila* might, eventually, be transferable to humans.

This would not be easy. One of the reasons *Drosophila* had become such an important organism in genetics research is that is has only four chromosomes. Human beings were known to have many more, all crowded into a tiny space, making microscopic investigation exceedingly difficult. In 1951, when Hsu began his work in Galveston—thirty-six years after T. H. Morgan and his colleagues published *The Mechanism of Mendelian Inheritance,* two years before James Watson and Francis Crick published their paper on the molecular configuration of DNA—the exact chromosome number for *Homo sapiens* had not yet been firmly established. Between 1890 and 1950 many cytogeneticists had tackled the question, but their estimates varied widely; some said eight, others fifty. After 1930 an estimate made by the highly respected American cytologist T. S. Painter—forty-eight—was taken to be likely, but not 100 percent certain.[8] Not surprisingly, during the interwar years most genetics research programs were focused on organisms like the fruit fly, from which it was far easier to get concrete and verifiable results.

However, in 1951 Hsu had no choice but to work on human chromosomes; he needed a job and this was the only one he had been offered. The head of the lab wanted him to figure out how to photograph—even film—the chromosomes in human tumor cells so as to understand why the cells divided so much faster and so much more often than cells in normal human tissue. During his first few months in the lab, Hsu learned basic human cytogenetics: how to set up cultures of human cells (cells have to be dividing—which is the way cell cultures grow—for chromosomes to be visible), how to take microphotographs as the cells divided, and how to make time-lapse films.

What happened next is best recounted in his own words:

> I liked the laboratory . . . but I was nostalgic, yearning for a return to *Drosophila.* Then came a miracle. The laboratory had received a few samples of fetal tissues from therapeutic abortions. We set up all sorts of [fetal] tissues for culture . . . I really did not have definite experiments to run, but I set up some skin and spleen cultures anyway. Then I fixed [killed] some of the cultures and stained them . . . I really was not looking for anything in particular, but I thought I might be able to see lymphopoesis [the division of white blood cells, important in the study of leukemia] *in vitro.* I could not believe my eyes when I saw some beautifully scattered chromosomes in these cells. I did not tell anyone, took a walk around the building, went to the coffee shop, and then returned to the lab. *The beautiful chromosomes in those splenic cultures were still there; I knew they were real.*[9]

Instead of being crowded and interwoven, the chromosomes in that particular set of miraculous slides were large and separated; Hsu could not only count them, he could also distinguish several different shapes.

Hsu soon learned that a technician in the laboratory had accidentally washed that culture of fetal spleen cells in a hypotonic solution (one less salty than the solution usually used). This had caused the chromosomes to take on water and swell, and that was why they were so easy to see under the microscope. Hsu tried using a hypotonic wash with cultures of other human tissues and with cultures of cells from other animals: it worked every time. Four months had passed since his original discovery; there was no way to tell who had made the original mistake, and no one in the lab was willing to own up to it. Hsu and the lab director were elated; they knew exactly how useful this small change in laboratory routine would turn out to be. "I would love to give a peck on the cheek," Hsu wrote in his autobiography twenty-five years after the event, "to that young lady who made an important contribution to cytogenetics."[10]

Hsu published a note about the new technique within a few months; a year later he and the lab director published a longer, more detailed paper.[11] Both papers were sent to the *Journal of Heredity,* a prominent professional publication read by geneticists, not by physicians. However, the laboratory director lost interest in pursuing cancer through the study of chromosomes, and Hsu had to drop his investigations. He did not return to cytogenetics until 1959, when he was finally able to set up his own laboratory.

By that time human cytogenetics had been completely transformed by the hypotonic technique, and by the work of two plant geneticists working in Sweden—a Swede, Albert Levan, and an Indonesian, Joe-Hin Tijo—who had developed an interest in human cancers. In the 1940s Levan had become expert at figuring out how various chemical agents caused plant cells to reproduce too rapidly, and in the early 1950s he turned his attention to the same problem in mammalian cells. In 1955 he and Tijo were examining the nuclei of cells from a culture of fetal lung tissue (fetal tissue was preferred for chromosomal studies because fetal cells divide rapidly and frequently). They prepared the tissue by pretreating it with colchicine (a chemical that they knew made plant chromosomes swell) and then washing it in a hypotonic solution. The result was excellent; the chromosomes could be seen very clearly; so they decided to try counting. Over and over again, they found a pair of sex chromosomes and twenty-two other pairs, making a total of forty-six, not forty-eight. Though they were excited by this surprising result, in the rush to get a paper out they reported their results tentatively. Within a few months, however, other researchers had replicated and confirmed their count in a paper published in *Nature,* perhaps the world's most widely read scientific journal.[12]

Several cytologists who had once shunned work with human chromosomes were now drawn into the field. If human chromosomes could be counted, these scientists reasoned, then they could also be characterized. And if they could be characterized in normal cells, then they could also be characterized in abnormal cells, particularly cancer cells. And if abnormal chromosomes could be found in abnormally reproducing cancer cells, then perhaps the cause—and the cure—for cancer could be found, right there, under the microscope. The prospect was tantalizing.

By 1960, just four years after Levan and Tijo first announced the correct count, a consensus conference was called to develop a standard nomenclature for describing chromosomes, since each of the research teams that were carefully observing their number and trying to characterize their shape had developed a different way of signifying what it was seeing under the microscope. The conference was held at the University of Colorado Medical School in Denver. There were only a dozen or so participants, representing laboratories in Sweden, Japan, France, Scotland, and the United States. Three geneticists who had not worked on human cells were also invited to moderate what were expected to be inevitable disputes about whose naming system worked best. The system developed at the Denver Congress is still used today, with just a few modifications. Each sex-determining chromosome was given a letter (X, Y), and each of the other pairs (autosomes) was given a number (1–22) and a description based on its size and form.

The Denver conference was paid for by the American Cancer Society. In the first decade of human cytogenetics, the gleam in the eye of researchers came from the often-expressed hope that knowledge of human chromosomes would lead to a cure for cancer. Research

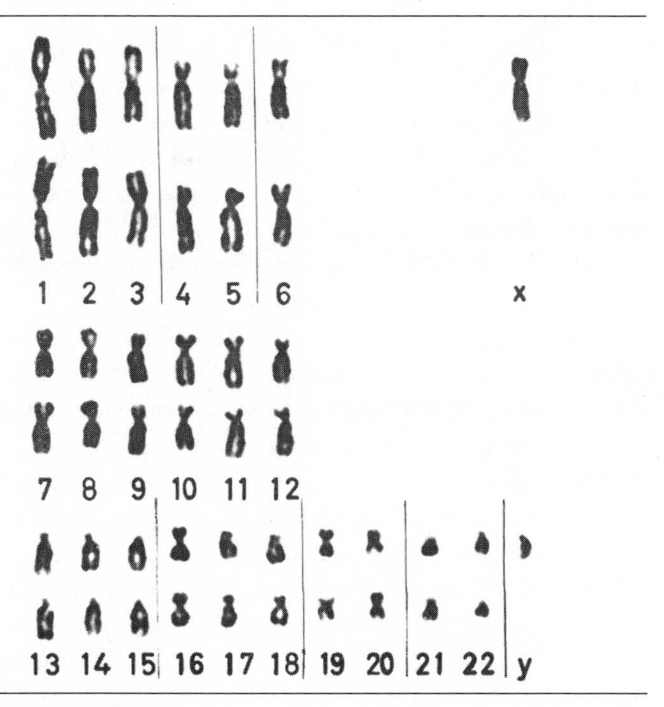

The Denver nomenclature.

From Denver Report, "A proposed standard system of nomenclature of human mitotic chromosomes," *Heredity* 51 (1960): 219. With the permission of Nature Publishing Group.

cytologists hoped that learning how chromosomes behaved when they were dividing abnormally would make it possible to figure out how to make them stop dividing. At the time, the notion that the very same knowledge might be used for genetic testing was not apparent to cytogeneticists.

The Clinician-Researchers

However, by 1966, when the rapid pace of discovery in human cytogenetics required that another consensus conference be called, the sponsoring agency was no longer the American Cancer Society, but rather the March of Dimes/Birth Defects Foundation. This foundation had been established in the 1930s to care for victims of polio. After it assisted in the development of a vaccine that would prevent future outbreaks of this disease, it turned its attention, and its considerable endowment, to fighting other conditions—like birth defects and prematurity—that damage the health of babies and children. The idea that human cytogenetics might unravel the mystery of congenital diseases (in addition to the mystery of cancer) emerged out of clever microscopic detective work done by a dozen or so adept clinician-researchers during the late 1950s.

Jérôme Lejeune was one of them. A medical student in Paris just after the end of World War II, Lejeune had become a clinical assistant to a pediatrician named Raymond Turpin. Turpin had a longstanding interest in the care of children with the disease then called mongolism, now known as Down syndrome. After Lejeune had finished his medical studies and his military service, he returned to Turpin's laboratory with the intention of working toward a Ph.D. Turpin suggested that Lejeune take over his Down patients, and the younger man began to plan dissertation research that would focus on this disease. His plans were constrained, however, by the fact that Turpin's laboratory had almost no research equipment beyond a table and a faucet, not even a microscope.

Making the best of his limited resources, Lejeune first turned to dermatoglyphics, the study of palm prints, an enterprise that required the use of statistics but no laboratory equipment. Down children were known to have unusual palm prints. Lejeune determined, first, that the unusual aspects of his patients' palm prints were consistently similar enough to be diagnostic of the disease, and second, that the unusual characteristics were also found more frequently in the palm prints of their relatives than in those of randomly selected individuals. Another line of dermatoglyphic research suggested that some of the lower primates had palm prints that bore some resemblance to the Down "stigmata."[13]

Lejeune did not pursue this line of research for very long. In 1954 Turpin received some funding to investigate the mutagenic impact of atomic bomb fallout; everyone in his laboratory, including Lejeune, was required to participate in this research. Four years passed before Lejeune was able to return to his Down syndrome work, but they were not wasted years, because during them he became an expert human cytogeneticist. The dermatoglyphic research had given him an idea: if there were several phenotypic traits that were so unusual as to be diagnostic for Down syndrome (distinctive palm prints, distinctive shape of the eyes, distinctive abnormality of the little finger, mental retardation), then the disease was not likely to be the result of a mutation in one gene but must be an abnormality of a whole chromosome.

As soon as he was able, Lejeune put the skills he had learned on the mutagenesis project to work on the cells of Down children. He removed tiny bits of tissue from inside the cheeks of some of his patients and, with the assistance of a colleague who had learned human tissue culture techniques (Marthe Gautier, a cardiologist who was head of the clinic for Down children), he fixed some of the cultured cells, washed them in hypotonic solution, and looked at them under the microscope.

As so often happens in the history of science, an observation with far-reaching consequences can be stated in very few words. The opening sentence of the paper that Lejeune subsequently published says it all: "The study of mitosis in cultured fibroblasts from nine mongoloid children . . . has allowed us to see, regularly, the presence of forty-seven chromosomes."[14] Forty-seven, not forty-six: there was an extra chromosome in people with Down syndrome. Thus in 1959 Down syndrome became the first form of mental retardation to be linked to a chromosomal abnormality. Within weeks of publication of this finding, it was confirmed by other teams of cytogeneticists; within months, in tacit acknowledgment of the finding's significance, other researchers were claiming to have found it first (although none of the claimants ever succeeded in displacing Lejeune's priority).

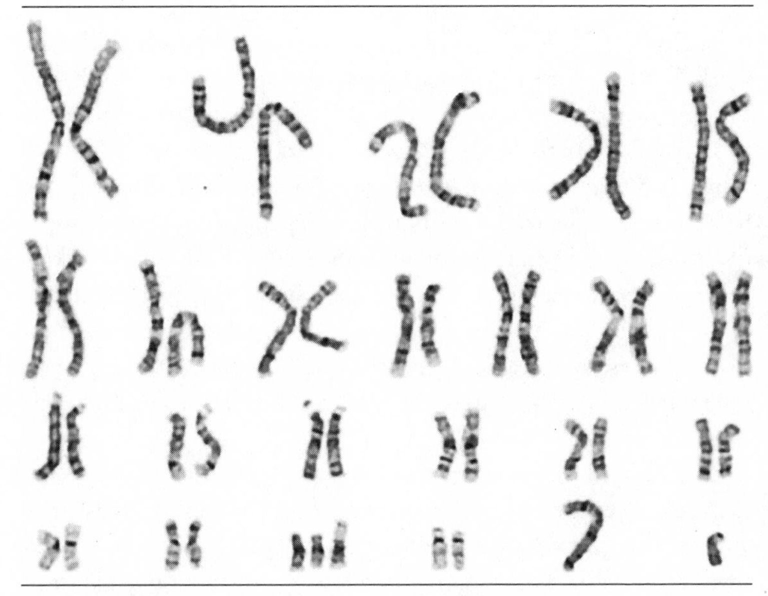

Karyotype of a male with Down syndrome, which demonstrates three copies of chromosome 21.

Courtesy of Nancy B. Spinner, Ph.D., University of Pennsylvania School of Medicine and Children's Hospital of Philadelphia.

The discovery of the extra chromosome linked to Down syndrome brought Lejeune both lasting prestige and lasting heartache. He was one of the cytogeneticists invited to the Denver conference where the supernumerary, or extra, chromosome was christened with the number twenty-one. The University of Paris awarded him a Ph.D. and, a few years later, a professorship in its faculty of medicine. By the mid-1960s Lejeune had become the head of his own laboratory, much more amply funded than Turpin's had ever been. Nonetheless, despite decades of systematic attempts, the discovery that many people with Down syndrome have what became known as "a trisomy of the twenty-first chromosome" did not lead to a cure for the condition, as Lejeune dearly hoped it would. Rather, the discovery led to the widespread use of amniocentesis to prevent the birth of babies with that disabling condition, a result that Lejeune would quickly come to abhor.

Enter a Neurophysiologist

Although Down syndrome is now the condition most often diagnosed via prenatal screening, it was not the first condition on which amniocentesis was put to use: sex was. Lejeune could see the extra twenty-first chromosome when he looked at tissues from his patients, but in 1959 he could not have seen it in tissues that came from fetuses because the fetal cells that float in amniotic fluid have to be cultured in order for a sufficient number of them to be caught just at the moment when the chromosomes are visible. Researchers were trying to develop a medium in which those cells would grow, but no one had yet succeeded. Someone had, however, discovered how cells could be used to determine fetal sex without having to look at the chromosomes: that person was a Canadian neurophysiologist named Murray Barr.

Barr was born on a modest farm in western Ontario in 1902 and died, eighty-seven years later, just fifteen miles from his birthplace. Educated at the Medical School of the University of Western Ontario, he spent almost his entire career (except for six years in the Royal Canadian Air Force during World War II) as a faculty member of his alma mater. Barr developed an interest in neurophysiology while in medical school, and after graduating he sought out laboratory training in the techniques of neurocytology (the study of nerve cells), in the hope that he might someday get a Ph.D. This hope was never realized (the war intervened), but lack of the degree did not prevent Barr from establishing an active research program at the university after being deactivated from military service in the summer of 1945.

His research program was straightforward and focused on the changes that occur in nerve cells when they have been active for long periods of time. His investigation involved subjecting one nerve of a research animal (he used cats) to electrical stimulation for hours on end, then killing the animal, dissecting out the nerves, preparing them properly, and looking at them under the microscope. Most of the results were unsurprising: after so many hours of activity some cellular bodies had become enlarged while others had shifted position. But one result was exceptionally surprising. Here is an abbreviated version of Barr's account:

> As the work progressed, neurons from one animal were discovered to be unlike those seen previously because the nuclear satellite was missing. Then after an interval, those of a second animal were similarly found to be different from the others. The discrepancy remained a puzzle for a week or so, when it occurred to me to check the sex of the animals . . . On checking the records the two cats without the nuclear satellite were found to be male and the others female.[15]

This was a serendipitous revelation: not what Barr was looking for, but exciting nonetheless. Nerve cells from female cats had a cellular body that was not found in nerve cells from male cats. Barr checked cells from other tissues of the same animals: females had the nuclear satellite; males did not. He then checked for its presence in human tissues preserved on slides in pathology labs: women had it; men did not. Barr could see this strange structure (or its absence) in resting cells, ones that were not dividing. He had discovered that you did not need to see chromosomes in order to determine sex.[16]

This was big biological and medical news: one short paper went to *Nature,* another to the *Lancet.* Like Lejeune's results, Barr's could be summarized in a sentence: "It appears not to be generally known, however, that the sex of a somatic cell . . . may be detected with no more elaborate equipment than a compound microscope following staining of the tissue by the routine . . . method."[17] Taken together, Lejeune's and Barr's straightforward findings would turn out to have far-reaching implications for prenatal diagnosis.

Barr soon dropped his experiments on cats in favor of exploring the diagnostic implications of what he called "sex chromatin" (later it was also known as the Barr body, a term Barr himself never used). In 1950 human cytogenetics was still in its pre-hypotonic dark ages. Now that Barr had found a relatively easy way to figure out the chromosomal sex of a person whose anatomical sex was anomalous, papers soon began to pour out of his lab about the chromosomal sex of hermaphrodites, transvestites, women with Turner's syndrome, and men with Klinefelter's syndrome.[18]

Barr never took his discovery into the realm of prenatal diagnosis. By the 1960s, when the hypotonic revolution in cytogenetics made it possible to diagnose chromosomal sex abnormalities by looking directly at the X and Y chromosomes, Barr had given up his work on sex chromatin and returned to his first love, neurophysiology, eventually coauthoring a very popular textbook on the subject and supervising all of its many editions.

In the hands of other scientists, however, the technique Barr had developed was the first to be used in attempts to diagnose a fetal condition prenatally. Not an obstetrician, he himself had never thought of initiating such investigations, though he welcomed them. His unwitting contribution to the development of prenatal diagnosis was honored by many awards (including one from the March of Dimes/Birth Defects Foundation). Throughout his life he expressed much pride in what he and his colleagues had accomplished in the early 1950s and, unlike Lejeune, no regrets at all about the ways in which others had subsequently extended his discovery.

Putting the Pieces Together

In 1953 Barr had discovered that sex chromatin analysis could be easily performed on human cells scraped from the inner lining of people's cheeks, the same easily obtainable cells that Lejeune would later use to investigate the chromosomes of Down children. Two years later Barr published a paper, again in the *Lancet,* in which he reported having tested the inner cheek cells of 140 subjects, 13 of whom were infants less than a week old.[19] Cheek cells are fibroblasts, just like the fetal cells that float in amniotic fluid. If sex chromatin analysis could be performed on fibroblasts from newborns, couldn't it also be performed on fetal cells? Could those floating fetal fibroblasts answer the age-old question, the one that doctors could never answer confidently, "Is it a boy or a girl?"

In the five months between December 1955 and April 1956, four sets of researchers, from very different parts of the world, announced that, yes, sex chromatin could indeed be found in fetal cells floating in amniotic fluid. Two of these researchers, Povl Riis and Fritz Fuchs, were Danish, affiliated with major hospitals in Copenhagen and also with the medical school of the University of Copenhagen. Three others were Europeans who had immigrated to Israel and were employed either by the Weizmann Institute in Rehovoth or by the Hadassah University Hospital in Jerusalem: Leo Sachs, David Serr, and Mathilde Danon (later Mathilde Krim). Three more, E. L. Makowski, K. A. Prem, and I. H. Kaiser, were on the staff of the medical school of the University of Minnesota in Minneapolis. The last to publish, and the only researcher who worked alone, was Landrum Shettles, then a gynecologist on the staff of Columbia University in New York. Barr had included infants in his study just to show that sex chromatin could be found at all stages of an individual's life cycle, but to people interested in obstetric research his discovery had another, greater significance.[20]

The Israeli group first tested the hypothesis the easy way: collecting samples of amniotic fluid in hospital delivery rooms, centrifuging the fluid, fixing and staining the cellular material, looking for sex chromatin under the microscope, and then checking their findings against the sex of the baby once it was born. The other three groups obtained amniotic fluid the hard way, by amniotic tap. All came up with the same result: samples of amniotic fluid, even those taken as early as twelve weeks into the pregnancy, revealed the correct sex of the fetus almost 100 percent of the time.

This was an astonishing result, but at first glance it did not appear to have much practical application, especially if you were obtaining the information just an hour or two before a baby was born. Amniocentesis was too risky to use, as the Copenhagen group put it, "just to satisfy parental curiosity." Maybe, they conjectured, "If the [amniocentesis] results are confirmed in animals . . . it might become of great significance in veterinary practice." But in humans, the only application any of the obstetricians could imagine was in sex-linked disease. "Regarding the clinical value of a correct prenatal diagnosis of sex," the Jerusalem group suggested, "the most obvious application would be in cases of certain sex-linked abnormalities."[21]

The sex-linked abnormality that obstetricians see most frequently in their patients is hemophilia: virtually all hemophiliacs are males, and their pedigrees reveal that the familial incidence of the disease can be traced only through their mothers' families, meaning that the condition must be a Mendelian recessive, carried on the X chromosome. Although males have the disease, females carry the recessive gene that produces it; every time a woman who is a carrier becomes pregnant with a male fetus there is a 50 percent chance that the baby will have hemophilia.

Not long after the Danish researchers had published their paper about fetal sex determination, Fuchs, the obstetrician of the pair (Riis was a cytologist), delivered a baby that hemorrhaged and died within a few hours. Many years afterward he could still recall the case vividly: "While we were talking to the mother, she happened to say that she was a known carrier of hemophilia. So I said to her, 'I don't know whether the hemophilia has killed your baby this time, but if you want me to tell you whether or not the baby will be in danger in a future pregnancy, I am willing to do the amniocentesis.'"[22]

Two years later the woman became pregnant again and took Fuchs up on his offer. The paper that Fuchs and Riis published in 1960, after they had done the amniocentesis and stained the fetal cells, tells us a bit more about her situation: "Her first pregnancy in 1951 ended with the birth of a living girl . . . In 1954 she became pregnant again, but this time she requested and was granted a legal interruption of pregnancy on eugenic grounds. Her third pregnancy was in 1957, and this time she decided to carry it through."[23] (This was the pregnancy that ended with a male infant who hemorrhaged and died.)

In January 1959, when the woman returned to the clinic for amniocentesis, she had made up her mind, as Fuchs and Riis put it, that "she would not continue her pregnancy if there was a risk of bearing a boy with hemophilia."[24] So Fuchs performed an amniotic tap, and Riis stained the floating cells. Sex chromatin was clearly present: the fetus was female. The woman proceeded with the pregnancy. Although she went into labor two months early, the baby girl, after spending three months in an incubator, went home from the hospital in good health.

The paper in which Fuchs and Riis published their account of this case has an apt and significant title: "Antenatal Determination of

Foetal Sex in Prevention of Hereditary Diseases." Fritz Fuchs was not the first person to do an amniotic tap, nor was Povl Riis the first cytologist to stain fetal cells to determine the sex of a fetus. The two physicians were, however, the first to publish a case report in which an amniotic tap and a primitive form of karyotyping were used to *prevent the birth* of a baby who might have a genetic disease. Their patient, a known carrier of hemophilia, was therefore the first woman to subject herself and her fetus to prenatal diagnosis. To put the matter another way, Fuchs and Riis were the first clinicians who put together the crucial but heretofore separate components of the technological system we now call prenatal diagnosis: they had obtained a sample of fetal tissue and subjected that tissue to a test to determine the sex of the fetus, and they would have been prepared—had the fetus been male—to arrange for termination of the pregnancy.

The Last Essential Component

Abortion, termination of the pregnancy, was the necessary step without which neither the amniotic tap nor the subsequent cytological workup could have achieved its preventive goal. Fuchs and Riis may not have been the first physicians to attempt prenatal diagnosis for a genetic disease, but they were certainly the first to publish an account of what they had done. In 1960 Denmark was one of the few countries in the world in which such an abortion was explicitly sanctioned by law. Had other physicians—in New York or London or Minneapolis or Jerusalem—helped a patient get an abortion after a similar diagnosis and gone public with what they had done, they would have risked losing their licenses, being fined, jailed, or prohibited, on pain of the other three, from doing it again.

Fuchs and Riis knew this very well. Early in their paper they describe and quote from the abortion law that the Danish parliament had passed in 1938 and had modified somewhat in 1956. The law specified that permission for a legal abortion could be granted "if there is a close risk that the child, due to inherited characteristics . . . may come to suffer from . . . severe and non-curable abnormality or physical disease."[25] The law also prescribed a procedure to be followed in these cases: the woman requesting an abortion had to apply to a government-supported social work agency called the Mother's Aid Institution; the agency had to contact the Institute of Human Genetics at the University of Copenhagen for verification of the risk; finally, a board consisting of a psychiatrist, a gynecologist, and a social worker had to give its permission. These particular abortions were referred to, by Fuchs and Riis and many of their contemporaries, as "eugenic" abortions because, in Denmark and the rest of Scandinavia, "eugenic" denoted "caring for the welfare of mothers and babies," not "involuntary sterilization and euthanasia."

The extent to which Danish society approved of eugenic abortion is made clear in the second case that Fuchs and Riis reported in their 1960 paper. This patient had had an illegal abortion in 1958 when she was unmarried. "The present pregnancy," Fuchs and Riis report, "was desired, but in the middle of the fourth month she was told she was a conductor of hæmophilia." Her attending physician had guessed at the possibility: three uncles and one of her cousins were hemophiliacs, and her own blood, when tested, took a fairly long time to coagulate. The physician sent her to Fuchs, who applied to the Mother's Aid Institution for permission to do the abortion if the fetus turned out to be male. The institution made a very interesting ruling: "The Board of the Mother's Aid Institution ruled that interruption of pregnancy would be granted if the foetus was masculine, *and added that interruption would have been granted without conditions if the test had not been possible.*"[26] In Denmark, as well as in the rest of Scandinavia, abortions to prevent the birth of a child who would be likely (not certain, just likely) to have a serious disease or a disabling condition—eugenic abortions—had been legally, politically, and morally acceptable since the late 1930s. Prenatal diagnosis had the potential (illustrated perfectly in this case) of *reducing* the number of eugenic abortions performed. If prenatal diagnosis had not been available, the likelihood that Fuchs and Riis's second patient's pregnancy would have been terminated was 100 percent. With prenatal diagnosis the likelihood was reduced to 50 percent; if the fetus was female, there would be no disability to prevent.

Thus, from the perspective of the physicians who first developed prenatal diagnosis, it was both a preventive and a pronatalist technological system: preventive because it avoided the birth of a child who was doomed to suffer; pronatalist because it reduced the number of eugenic abortions being performed, while at the same time encouraging at-risk couples to have children.

Many years later Fritz Fuchs could still recall what that first diagnosis via amniocentesis had meant to him and to his first patient, the one who discovered that the fetus she was carrying was a girl and who, as a result, decided to carry her pregnancy to term. "They lived in my neighborhood," he said, "so I could see the mother very often walk with the pram and the baby. She was very happy that she could have the baby. And so was I."[27] The first provider of prenatal diagnosis was a physician who wanted to help the carrier of a genetic disease *have* a baby; the first patient was a woman who, absent the test, would have probably terminated her pregnancy. The goal, for both, was preventing the birth of a person who was very likely to have lived a short life, full of suffering.

Innovation and Diffusion

Beginnings count—and so do motivations, because motivations determine the goals of technological systems and the goals, in turn, shape the content. By 1960 both the goals and the content of prenatal diagnosis had been articulated. The content was a technique for obtaining and observing the tissue of living fetuses, one or more diagnostic tests that could be applied to those tissues, and a medico-legal system that allowed for the termination of afflicted pregnancies. The motivations of the men who developed the system were also clear: to allow people who were at risk of passing on a serious disease or disability to have babies without fear and, at the same time, to prevent the birth of children who were doomed to lives of suffering.

Over the course of the next twenty-five years demand for prenatal diagnosis began to grow in many countries. And as demand grew a number of techniques were developed that embellished the system in different ways, without changing its basic content or its goals.

Much of the demand came from medical geneticists and their patients. During the 1940s and 1950s only a few researcher-clinicians were interested in genetic disease, but as their interests became known, their patient base began to grow. Two kinds of patients might be referred to them: women who had already given birth to a child with a known genetic disease (such as hemophilia) and women who had not yet had children but who suspected that either they or their partners were carriers of some genetic abnormality. All these patients and their partners were referred to medical geneticists with one central question on their minds: How can I have children who will be free of this disease?

In those years there were no easy answers to that question. Patients of the first type were generally told that their risk of having another afflicted child was very high and that they should avoid another pregnancy, by whatever means seemed best. Patients of the second type fared little better, even if they were able to provide fairly complete family histories (as only a few could). All the physician could do was to calculate probabilities and provide advice about how best to avoid pregnancy.

Plagued by an ambivalence that was complicated by uncertainty, some at-risk couples ignored their physician's advice, playing a form of genetic roulette; most of the time, these couples did not stay abstinent, did not have themselves sterilized, and did not seek out artificial insemination or adoption. Their desire to bear children to whom they would be biologically related apparently trumped what they may have considered a small risk of having a seriously disabled child—until their first disabled child was born, after which they found ways to avoid subsequent pregnancies. The birth-control pill did not come on the market until 1960; as a consequence, medical geneticists frequently found themselves assisting their patients with the difficult task (outside Scandinavia) of obtaining a safe but only quasi-legal abortion.

Prenatal diagnosis had the potential for reducing all those frustrations. It could reduce the uncertainty in the risk estimates by at least providing a reliable diagnosis for some diseases in some pregnancies. As a consequence, some at-risk patients had more assurance of being able to bring healthy children into the world—and fewer occasions to seek abortions. The emotional pain of terminating a pregnancy, the professional pain of not being able to give a patient a clear diagnosis, the anxiety of waiting for a group of strangers to make a decision that would affect a whole family's future: all of that could be avoided through amniocentesis. Physicians who had not been able to give their patients firm answers hoped that amniocentesis would now allow them to do so. Patients hoped that it would release them from the fear of having children who would suffer—and from the fear, with every pregnancy, that an abortion could not be obtained. Demand for the prenatal diagnosis afforded by amniocentesis increased in the years following 1960—a beacon of hope in a medical domain that had never before had one.

As a consequence, improvements in the basic system came fairly quickly. Making a blind amniotic tap during the second trimester of a pregnancy was a risky undertaking because of the possibility of hitting some part of the fetus with the needle. Indeed, the second of Fuchs and Riis's patients suffered a miscarriage, which might well have been attributable to the tap they had performed. Waiting until the third trimester, when it was fairly easy to locate the fetus by probing the pregnant woman's abdomen, was not an acceptable option; third-trimester abortions were both medically and emotionally difficult for everyone involved. X-raying, another possibility, was both very difficult and potentially dangerous to the fetus. By the mid-1960s, however, a different observational technology was available, at least in some large metropolitan hospitals.

In the late 1950s a Scottish obstetrician and gynecologist, Ian Donald, began experimenting with an observational system that he hoped would improve his diagnostic abilities: ultrasound. Donald hoped that ultrasound waves would enable him to see internal anomalies in his patients, either when X-rays were not useful (if, for example, fat tissue was in the way) or when they were inadvisable (if the patient was pregnant). His first success with ultrasound visualization came in 1957, when he used it to correctly diagnose an easily removable ovarian cyst in a patient who had been thought to have an inoperable stomach cancer. In the next few years Donald achieved dozens of other diagnostic triumphs with ultrasound and published his case reports in prominent journals. "As soon as we got rid of the backroom attitude," he later wrote, referring to the skepticism of some of his colleagues, ". . . and brought our apparatus fully into the Department with an inexhaustible supply of living patients with fascinating clinical problems, we were able to get ahead really fast. Any new technique becomes more attractive if its clinical usefulness can be demonstrated without harm, indignity or discomfort to the patient."[28]

Obstetric ultrasound became popular very quickly, partly because of diagnostic successes achieved by its earliest adopters, partly because it was not invasive ("without harm, indignity or discomfort," as Donald aptly said), and partly because it allowed almost instant visualization, as no film had to be developed. In addition, the equipment was relatively inexpensive, perhaps not for solo practitioners, but certainly for hospital departments. By 1965 *Life* magazine was reporting: that "doctors use . . . sound to keep watch over their unborn patients," and put a photograph of a pregnant woman undergoing an ultrasound scan—with a cross-section of her fetus's head visible on the screen—on its front cover.[29]

Some of the physicians who were performing amniocentesis during the 1960s were also experimenting with ultrasound in an effort to avoid the risk of causing harm to the fetus with their needles. By the time another decade had passed, ultrasound guidance had become a standard component of the technological system. By the mid-1970s improved equipment made it possible to see not just the fetus but also the placenta. During those years ultrasound equipment also became available to more physicians, and radiologists had become more expert in its use. A survey undertaken in 1974 of fifty-five North American hospitals in which amniocentesis was regularly being performed revealed that thirty-five of them were either always or sometimes using ultrasound to guide the procedure; most of the remaining hospitals reported that they would use ultrasound guidance for amniocentesis if they had better, or more accessible, equipment.[30]

In addition to improving the safety of amniocentesis, medical geneticists also wanted to expand the range of conditions that it could be used to diagnose. All sex chromatin analysis could diagnose was the sex of the fetus—and that was medically useful only in a limited number of sex-linked serious conditions like hemophilia and muscular dystrophy. From the very beginning, Riis and Fuchs had believed that more conditions could be diagnosed prenatally if cytogenetic analysis became possible, that is, if some method could be found for culturing the fetal cells floating in amniotic fluid so that the chromosomes (not the chromatin) could be seen under the microscope. In 1962 Fuchs and one of his colleagues succeeded in keeping some amniotic fluid cells alive in their laboratory, but it

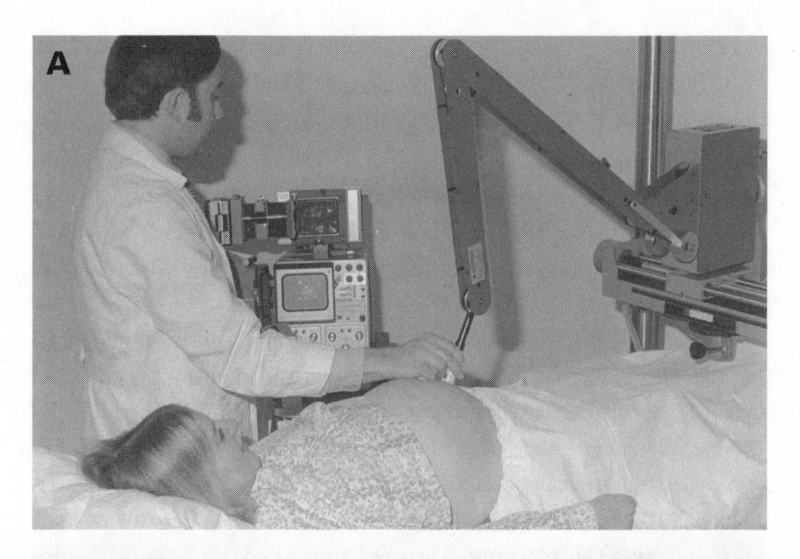

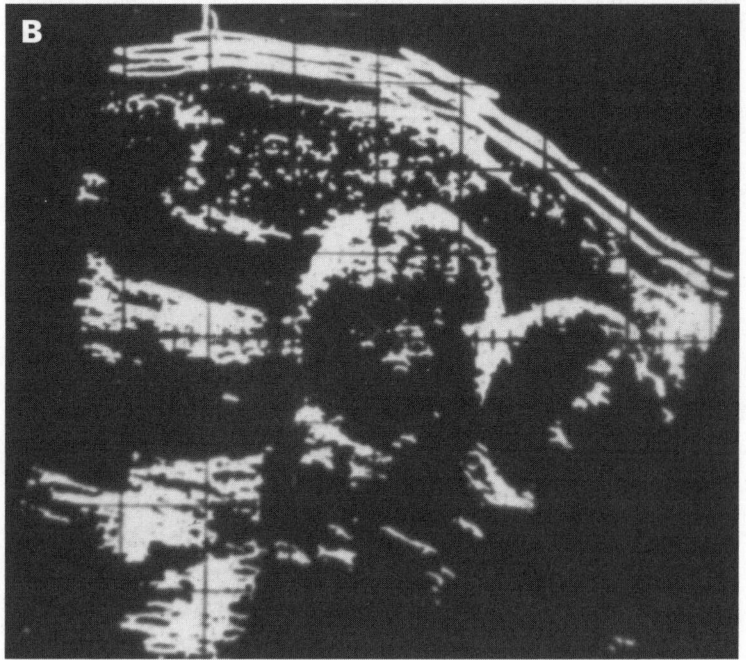

(A) Dr. Barry Goldberg performing an obstetrical ultrasound examination in 1971 when he was affiliated with Episcopal Hospital in Philadelphia. (B) The ultrasound image produced by this examination was of a twin pregnancy; both fetal heads can be seen, as well as the placenta.

Courtesy of Barry B. Goldberg, M.D., Thomas Jefferson University Hospital, Philadelphia.

took so many weeks to get a culture large enough for a few mitotic divisions to be visible under the microscope that the pregnancy would have been too advanced for a legal abortion if an unfavorable diagnosis had been made.[31]

The culturing problem was finally solved in 1966 by two American physicians who had a special interest in the cytogenetics of mental retardation, Mark W. Steele and W. Roy Breg. In a paper published in the *Lancet,* Steele and Breg announced that they had found a way to grow enough cells for karyotyping in a reasonable period of time and that they had determined that the cells they grew were fetal and not maternal. "A reliable technique for culturing foetal amniotic-fluid cells," they concluded, ". . . would allow more practical genetic counseling of mothers with high risks of having children with chromosome abnormalities."[32]

They were absolutely right. Between 1967 and 1969 numerous papers appeared in the professional literature in which physicians reported the diagnosis of various chromosomal abnormalities that had been associated with mental retardation, not just trisomy 21, but also what were then called the D/D and D/G translocations. With the ability to diagnose these causes of mental retardation, the potential patient base widened significantly. Some chromosomal abnormalities were indeed familial, passed down from parent to child, but the vast majority were not (one of those inconvenient facts that made eugenic sterilization of the feebleminded unlikely to succeed). Trisomy 21 seemed to be a fresh mutation each time it appeared, which meant that the patient population at risk was all pregnant women, or, since the incidence of the condition appeared to increase with the age of the mother, at least all "elderly" pregnant women.

In addition, by 1970 several physicians had gone beyond cytogenetic analysis to undertake enzyme analysis of cultured fetal cells. They discovered that biochemical studies of amniotic cell cultures could be used to diagnose several serious inborn errors of metabolism. Pushing even further, some researchers succeeded in staining *un*cultured fetal cells so as to reveal the presence of granular materials that signaled what were then called storage diseases, such as Tay-Sachs.[33]

Even the amniotic fluid itself turned out to be informative. In 1972 a biochemist at the University of Edinburgh, D. J. H. Brock, discovered that elevated levels of a particular protein (alphafetoprotein, or AFP) could be found in amniotic fluid when a fetus had an open neural tube defect, such as spina bifida. Since these defects were then very common in Great Britain (almost 8 per 1,000 births in Northern Ireland and 5.6 per 1,000 in Scotland), and since the results could be obtained without culturing cells, the AFP test was quickly added to those already being done following amniocentesis. The neural tube defects were like Down syndrome in that only a small number of cases (about 2 percent) were familial; the rest appeared without a family history, which meant that all pregnancies were at risk.[34]

As the pool of at-risk patients expanded, so did the demand for prenatal diagnosis. Knowledgeable physicians estimated that amniocentesis for genetic diagnosis was performed roughly three hundred times in the United States between 1967 and 1971, but three years later, three thousand or so were performed in just one year.[35] In 1960, when Riis and Fuchs tried amniocentesis for the first time to determine sex, the pool of potential patients had been limited to women who were known carriers of a sex-linked disease. By the early 1970s, as the result of cell-culturing techniques and the development of biochemical tests for a variety of conditions, the potential patient pool had come to include every pregnant woman.

Experts in medical genetics, obstetrics, and public health agreed that the time had come to assess both the safety and the diagnostic accuracy of amniocentesis. By the early 1970s clinical trials were being organized in Canada, the United States, and Britain. The U.S. trial was organized by the National Institute of Child Health and

Human Development (NICHD), one of the constituents of the National Institutes of Health (NIH). Nine hospitals across the country that had staff physicians actively involved in related research and clinical applications were asked to participate. Every time a hospital analyzed amniotic fluid from a pregnancy, demographic data was gathered about the patient, and then a control patient with similar characteristics was sought, until a thousand subjects and a thousand controls had been registered. When a subject-patient received an unfavorable diagnosis for the fetus she was carrying and decided to terminate the pregnancy, diagnoses that had been made prenatally were confirmed by postmortem examination of the fetus. All the infants delivered by subject-patients and by control-patients were carefully examined at birth and reexamined one year later.

The study began in 1971 and ended in 1973, after which it took almost two years for the data to be analyzed and a final report released. The results were astonishing, even to those physicians who had had a great deal of experience with the procedure:

> There was no statistically significant difference between the two groups [of women] in rate of fetal loss (3.5% for the subjects, 3.2% for the controls) or incidence of complications of pregnancy or delivery. Newborn examination indicated no significant differences between the two groups [of infants] . . . and no evidence of physical injury resulting from amniocentesis. The two groups [of infants] did not differ significantly in physical, neurological, or developmental status at one year of age. Diagnostic accuracy was 99.4%.[36]

By the time the similar Canadian and British results were published a year later, many American insurance companies had decided to cover not only the costs of the procedure but also the costs, when chosen, of the subsequent abortion. By the end of the decade, Canadian and British obstetric clinics had decided to offer amniocentesis to some older pregnant women and to all women who had a familial history of any of the conditions that were then diagnosable. All of this signified that, by the end of the 1970s, prenatal diagnosis had become a standard part of prenatal care. This status was confirmed in 1977 and 1978, when several American women who were over thirty-five and had given birth to children with Down syndrome successfully sued their obstetricians for malpractice for failing to advise them that a prenatal test for the condition was available.[37]

Innovation continued at a brisk pace, sometimes to solve problems that had become apparent with increased use of the procedure, sometimes to extend diagnostic capabilities. When a test was developed that could detect AFP in a pregnant woman's blood toward the end of her first trimester, maternal AFP testing was added to the list of recommended tests on the blood of every expectant mother who came for prenatal care, not just in Britain but in other countries with high incidence of the condition, such as the United States.[38]

Ultrasound examination also changed markedly in the 1980s and 1990s. As demand increased, more companies entered the market, and as more companies entered the market, prices came down and the rate of innovation went up. Transducers (the part of the equipment that emits the sound waves) became easier to use, and images became easier to read. The first real-time ultrasound scanners, which produced moving images rather than still ones, were constructed in the mid-1960s and became commercially available a decade later. By 1980 they could be found in many hospitals and clinics; a decade later, as prices fell, individual practitioners were able to purchase them. Originally intended for use in guiding amniotic taps, the equipment became increasingly useful for direct diagnosis: first to date pregnancies by measurement of the fetus, then to see anatomical abnormalities, such as neural tubes that did not close, or brains that did not develop, or hearts with occluded valves, or—by the mid 1990s—the amount of fluid that was accumulating in the neck of the fetus, which points to a likelihood of Down syndrome.[39] The sex of a fetus could also be revealed by these improved ultrasound devices, and by the end of the 1980s physicians all over the world were offering ultrasound diagnosis of sex to couples who intended to terminate pregnancies unless the fetus was of the sex they preferred. The use of ultrasound followed by abortion of female fetuses became so pervasive in India—and the feminist outcry against it so loud—that the practice was officially outlawed by the Indian federal government in 1994, although it persists, virtually unabated, in India and also in China and several other Asian countries.[40]

Chorionic villus sampling (CVS) was introduced in the 1980s. Amniocentesis could not be done until the fourteenth or fifteenth week of a pregnancy; before then there was neither enough fluid nor enough living cells floating in the fluid for culturing. As a result, the abortions that sometimes followed were late-second-trimester abortions, which are both emotionally and medically difficult, not only for the patient and her partner but for the surgical team as well. In some jurisdictions these abortions remained illegal, even under reformed abortion laws; in others they could be subject to litigation; in still others permission was required from several layers of authorizing agencies. A test that could be done earlier in a pregnancy would solve this problem.

Several obstetricians recognized that the chorion might provide a solution. The chorion is one of the three membranes that surround the fetus. Like the other two membranes, chorionic cells come from the fetus, not from the mother. As the fetus grows, the chorion develops small projections (called villi) which grow larger where they come in contact with the wall of the uterus, eventually forming the placenta. Might it prove possible, by slipping a tiny microscope and a biopsy knife through the cervix, to cut off one or two of those villi without endangering either the mother or the pregnancy?

The answer to that question turned out to be a qualified yes; it could be done, but not safely, at least not until an extremely small fiber-optic probe was invented. In the late 1960s and early 1970s several Scandinavian physicians started trying CVS with instruments that were five to six millimeters in diameter, but the miscarriage rate that resulted was, in their view, far too high (all their research subjects were pregnant women who were already scheduled for first-trimester abortions). A team of Chinese physicians reported success in 1975, but Western experts did not believe their assertion that, using only a biopsy knife, it was possible to achieve a miscarriage rate of only three in one hundred. The Chinese team was only interested in diagnosing for sex; of the thirty induced abortions they reported in their paper, twenty-nine were of female fetuses, suggesting that the effort was a response to Chinese population policy and that the miscarriage rate was not of particular concern to the researchers. A combination of reasonable safety and reasonable accuracy was not achieved until the early 1980s, with instruments that had been reduced to 1.5 millimeters, by teams working in Hungary, Britain, France, and Italy.[41]

The British and French teams had a particular set of reasons to search for an early diagnostic technique. They were working with patients who were known carriers of one of the devastating blood diseases, sickle-cell disease or β-thalassemia. Some members of the research teams were pediatricians who were caring for children who had already been born to these women. Since the mid-1970s these pediatricians had been referring these mothers, when they became pregnant again, for fetoscopy, an exceedingly difficult surgical technique by which a sample of fetal blood was removed from the umbilical vein during the second trimester of pregnancy. Because these physicians knew the families with whom they were working, they knew that there was great pressure on the women to keep having children. They also knew that the women were so desperate to have disease-free children that they were willing to subject themselves to this extraordinary procedure and to undergo second-trimester abortion, despite the fact that their cultures prohibited abortion (many of the British patients came from Cyprus and Pakistan; many of the French patients, from North Africa). DNA markers for the genes that caused these two diseases were found in the late 1970s, which meant that direct analysis of the DNA from fetal cells could spare these women the fetoscopy—and if the cells could be obtained early enough, abortions could be performed before the pregnancies became obvious.

CVS caught on very quickly, although it has never quite supplanted amniocentesis as the most frequently used method for obtaining fetal tissue. CVS is a more difficult technique for physicians to learn, and as a consequence the induced-miscarriage rate has always been slightly higher for CVS than for amniocentesis. Several reports of malformed limbs following CVS appeared in the medical literature in the 1990s, and although large-scale studies failed to confirm the connection, many physicians and patients remain leery of the procedure. In addition, chorionic tissue is less informative than amniotic fluid, since there are some conditions (for example, open neural tube defects) that can best be diagnosed by biochemical analysis of amniotic fluid, not fetal cells.

Abortion Reform and Prenatal Diagnosis

Not at all coincidentally, the legal status of abortion changed completely during precisely the same years, 1960–1985, in which prenatal diagnosis was becoming a standard part of prenatal care. Women have been trying to end their pregnancies for thousands of years, in every culture that we know of. In many societies there have been experts—physicians, midwives, folk healers, pharmacists—who have assisted them. Nevertheless, abortion practices, although widespread, were often taboo. In many industrialized nations what had once been taboo was made illegal at about the same time—the middle decades of the nineteenth century—when biologists discovered that embryological development starts when the nuclei of an egg cell and a sperm cell merge.

Demand for abortion did not cease, but the fact that it was illegal drove it even further underground. Because what is clandestine is often unsafe, most of the Scandinavian countries legalized abortion, under very strictly controlled regulations, in the late 1930s as part of their effort to create programs that would enhance maternal health. As abortion laws changed between 1965 and 1985, what had once been legal and acceptable only in Scandinavia became both legal and acceptable in many developed countries of the West. What had been clandestine and unsafe for decades became legal, sanctioned, accessible, and therefore safe (safer than giving birth, as it turned out).

Abortion-reform movements of the early 1960s were generally led by professional people, physicians who were under increasing pressure from their patients to perform abortions or who were seeing the grim results of botched abortions in emergency rooms; lawyers who understood the involvement of so many ordinary people in an illegal activity as a threat to the rule of law; and clergymen who found themselves counseling people to break the law in this particular way. By the mid-1960s these professionals had been joined by increasingly angry feminists, large numbers of women who felt they were entitled to control the number and spacing of their pregnancies so that they could pursue education, employment, or professional careers.

All successful social movements—and abortion reform was indeed very successful—are galvanized by one or two crucial events, the kind that recruit more activists to the cause at the same time as they alter the receptivity of people with the power to make the desired reform. Both of the events that galvanized abortion reform happened in the early 1960s, and both had something to do with eugenic abortion. The first was what came to be called the thalidomide scandal; the second was an epidemic of rubella that swept through the United States between 1962 and 1967.

When thalidomide came on the market in Europe in the late 1950s, it was used as a sedative, and it was also given to pregnant women to minimize the discomfort of morning sickness. By the early 1960s European newspapers were carrying occasional stories about women who had taken the drug early in their pregnancies and had given birth to babies with missing or stunted limbs. Pregnant women who realized that they had unwittingly put their fetuses at risk by taking thalidomide began asking their physicians to perform abortions. Sometimes those requests were denied. One American woman who faced that situation allowed her case to be publicized. Sherri Finkbine, a television broadcaster in Phoenix, Arizona, had taken the drug in 1962, early in her fifth pregnancy. When she learned about the potential harm to her fetus, she asked her physician to perform an abortion, and he arranged to do so in a hospital with which he was affiliated. That women with the "right connections" could get such safe abortions in those years is another reason why doctors and lawyers thought reform was necessary.

In an effort to warn other mothers about thalidomide, Mrs. Finkbine told her story to a friend who was a newspaper reporter, but as soon as the story was published the hospital withdrew its permission for the operation. Within days the Finkbine case had attracted national and international media attention, which meant that when the Finkbines flew to Sweden to have the abortion done (as a legal eugenic abortion), their photographs appeared on the front pages of dozens of newspapers—as did the news that such abortions were legal in Scandinavia. If Arizona had adopted the model abortion statute proposed by the American Law Institute in 1959, Finkbine's abortion would have been completely legal—a fact that was broadcast widely by reform activists in the next few years.

The impact of the rubella epidemic in the United States was also widely publicized by reformers. Rubella, or German measles, is a relatively harmless disease of childhood. When a pregnant woman contracts the disease, the risk that she will deliver a neurologically impaired baby increases dramatically. A vaccine to immunize children against rubella was not available until 1969; consequently, when the epidemic began in the early part of the decade, many thousands of women were at risk. In the year 1964 alone, slightly more than twenty thousand babies in the United States were born with rubella-related deafness and blindness. Pressure on physicians to perform abortions increased considerably in those years, as did pressure on state legislatures to consider the financial implications of having to provide services for large numbers of disabled children and their families.

The result of all this mounting pressure was not only that abortion began to be legalized in many jurisdictions but also that fetal deformity began to be listed as one of the indications for a legal abortion. By 1973, when the United States Supreme Court ruled that women did not have to provide reasons for wanting an abortion, eighteen states had already passed abortion-reform measures that made legal abortion dependent on the length of the pregnancy, the approval of a board of experts, and the offering of an acceptable rationale: in most of those states a high likelihood of fetal deformity was considered acceptable, as it was in virtually all the jurisdictions—nations, counties, provinces—that reformed their abortion laws after 1965.[42]

Thus, at the very same time that the technological system of prenatal diagnosis was improving in safety and accuracy, the last crucial component of the system, abortion, became legal, safe, accessible, and affordable in most of the developed world.

Almost no one objected to prenatal diagnosis in these early days, except those who objected to abortion on any grounds whatsoever.

However, as the procedure began to diffuse, as it shifted from an experimental practice to a standard one, and as it was offered to more and more patients, objections became more frequent.

Some of those objections were part of the increasingly rancorous debate about abortion. Insofar as it is possible to judge by the editorials and letters to the editor in medical journals, by 1960 most physicians had become what Kristen Luker has called "loose constructionists" on the subject of abortion for fetal indications; that is, even in locales where most abortions were illegal, physicians were willing either to perform them or to look the other way when other physicians performed them if there was a high likelihood that the newborn would be severely ill or disabled.[43]

As legislatures began to pass laws that made the likelihood of fetal defects grounds for obtaining a legal abortion, and as accurate diagnostic tests for such conditions as Down syndrome and spina bifida were developed, opposition quickened both inside and outside the medical community. Members of the American Catholic hierarchy began telling their parishioners to stop raising money for the March of Dimes/Birth Defects Foundation. Pickets were established in front of hospitals that permitted their physicians to perform abortions for fetal indications. Some of the clinician-researchers who had developed component parts of the amniocentesis system—Jérôme Lejeune, who had discovered trisomy 21; William Lilley, who along with D. C. A. Bevis had helped perfect amniocentesis for Rh diagnosis; Ian Donald, who had created obstetric ultrasound—began, quite publicly, to express their dismay.

In 1969, upon being awarded a prize by the American Society of Human Genetics, Lejeune gave a lecture in which he warned his colleagues that "to kill or not to kill" was a question they could no longer avoid facing, even if their professional training had given them no reason to think they ever would. A decade later, still incensed, Lejeune testified before the United States Senate judiciary committee in favor of the notion that human life begins at conception; he also began signing petitions arguing that disabled newborns should be given every possible form of medical attention, even against the wishes of their parents. Like Albert Einstein before him, Jérôme Lejeune had come to regret one of the technological uses to which his scientific discovery had been put.[44]

Medical geneticists as well as many obstetricians and pediatricians were quick to counter such arguments. Physicians who counseled couples at risk of having children with a disease or disability saw amniocentesis as a way to reduce the number of abortions while also enabling the couples to have children without fear. "The patient greatly wants to have children and is willing to become pregnant again," said the team that reported the first prenatal diagnosis for Down syndrome in 1968, *"provided that the same diagnostic technique is applied to her future unborn babies."*[45] Physicians who were committed to the social welfare goal of improving the health of poor mothers and children believed that if amniocentesis reduced the number of mothers caring for profoundly disabled children, then the abortions that would result would be compassionate ones, entirely justified as a benefit to families. An ill child could undermine a mother's health and thereby undermine the health of her whole family. Physicians who cared for terribly ill and disabled infants and children, even those from affluent families, saw amniocentesis as a way to prevent suffering, not only the suffering of their patients, but also that of the patients' parents and siblings. "This is a heartening thought," one pediatrician wrote in 1966, since "the prevention of severe congenital malformations holds out more hope than their cure."[46]

Three different adjectives were used, over the years, to signify that these abortions for fetal indications were, somehow, different from other kinds of abortions—and most of them carried positive implications. "Eugenic" abortion was used, at first, in Scandinavia, to signify that these abortions were good for mothers and families.

When the word "eugenic" became taboo, "therapeutic" or "preventive" was substituted, to signify that in the absence of a cure, prevention was a good alternative therapy.

In addition to these physicians, many millions of women have voted with their feet in support of prenatal diagnosis; consenting to allow experimental trials, requesting the service when their attending physician has neglected to offer it; returning for prenatal diagnosis again and again; sometimes, in the United States, suing their physicians for failure to advise them about the existence of the tests. Virtually all women (and men too) who know themselves to be carriers of a serious familial disease and who have relatives who are afflicted want prenatal diagnosis so that they can bear disease-free children. Many parents of children with genetic or congenital diseases also support the testing regimens because they want their "healthy" children to be able to bear children, when the time comes, without fear. Prenatal diagnosis has, indeed, provided additional opportunities for pregnancies to be medicalized, but many women wanted precisely this form of medicalization, so that they could be assured, not that their children would be perfect, but that their children would not suffer from a devastating illness.

Opponents of prenatal diagnosis have frequently raised the specter of eugenics in their effort to gain adherents to their cause. "The idea of genetic therapeutic abortion is incorrect and dangerous," a German physician wrote to the *Lancet* in 1966, ". . . [and] differs little from the euthanasia programme practiced in Germany in Hitler's time." Forty years later a book reviewer made the same connection: "The search for a good birth, the search for the perfect baby, the search for a culture free of 'abnormal' people is now deeply entrenched in our medical community and society at large. Of course the last time we saw eugenics unleashed on a society as a whole was in Germany in the 30s and 40s." Even supporters of prenatal diagnosis bring up its historical connections with Nazi eugenics, if only to caution opponents that, to paraphrase the argument, we know enough about what happened then to make sure that nothing like it will ever happen again.[47]

They are all wrong. Prenatal diagnosis has almost nothing in common with eugenics, neither historically nor technologically. Eugenics was based on family pedigrees, which were themselves based on observations that could not conceivably be double-checked, since many of the people being described were dead. The genetic ideas that are at the core of prenatal diagnostic testing (for example, that the gene for hemophilia is on the X chromosome, or that people with three copies of the twenty-first chromosome have some degree of mental retardation) are based on causal connections or correlations that have been repeatedly verified over dozens of years.

Even more telling, none of the basic parts of the technological system of prenatal diagnosis was developed by eugenicists or with eugenic intentions: the amniotic tap was intended to help prevent stillbirths; karyotyping was developed by people who wanted to cure cancer; sex chromatin analysis was discovered by someone who was studying the neurophysiology of exhaustion. Furthermore, eugenicists never wanted to curtail their own breeding potential. No matter how they defined the object of their attentions (whether by skin color, or religion, or class, or health status), "the other" was always someone else, someone who was somehow "unfit," someone they thought should not be allowed to bear (or continue to bear) children. The technologies they chose reflected their goals. Early on, eugenicists chose to curtail the reproduction of those they deemed unfit by the technologies of sterilization or of sexual segregation; later, a few went further, choosing first the technologies of segregation and, subsequently, the technologies of mass murder.

The physicians and scientists, like Fritz Fuchs and Povl Riis, who first put together the component parts of prenatal diagnosis had completely different objectives from the eugenicists and hoped for very different outcomes. The people they wanted to help were potentially (sometimes even really) their neighbors, friends, patients, wives. Primarily, they wanted to reduce individual suffering: the pain of a boy with hemophilia, who bleeds internally every time he falls down; the burden of a mother who must neglect herself and her other children to care for a child who cannot learn to feed or dress herself; the pain of a sibling who knows he will have to commit his brother to an institution after his parents have died. Obstetricians and medical geneticists involved in the early days of prenatal diagnosis were particularly focused on the suffering of individuals and couples who feared they could not bear healthy children: the pain of a man who decides not to marry for fear that his children will have the same disease his brother has; the pain of a woman who aborts all her pregnancies for fear of having a child with the disability that embittered her father.

These hoped-for outcomes are fundamentally anti-eugenic. If their efforts were successful, at-risk couples and at-risk individuals—precisely the people the eugenicists might once have wanted to sterilize—would have more children. The early medical geneticists chose the components of their technological system carefully. They sought out fetal tissue, taking cognizance of the pedigrees and the phenotypes of the parents but not relying solely on such faulty predictive tools. They searched for safer ways to obtain or observe that tissue, developing less invasive procedures (like maternal blood screening to replace amniocentesis, or chorionic villus sampling to replace fetoscopy) so as to lower the risk of miscarriage. Realizing that they could not yet offer therapies or cures for most of the conditions they found, they helped their patients find safe abortions when abortions were still illegal, and they supported abortion-reform movements, sometimes financially, sometimes by lobbying, but most often simply by offering prenatal diagnosis to more and more patients.

In short, as a genetic test, prenatal diagnosis produces a result that is completely the opposite of what a eugenicist would have desired: it allows people who are genetically "at risk" to have as many children as they want. Furthermore, when prenatal diagnosis is understood as a technological system, its history demonstrates that it incorporates a novel stance on the morality of abortion: it asserts that the reduction of suffering can be understood as a fully ethical justification for the termination of pregnancies. Two kinds of suffering are being weighed in this calculus: the suffering of the potential person who will not be born and the suffering of the parents and siblings on whom that potential person would be chronically dependent. In that sense—specifically because of its ability to take the burden of lifelong care off the shoulders of mothers—the technological system of prenatal diagnosis also incorporated a feminist understanding of motherhood, as a choice that women could make without fear of a lifelong overwhelming burden.

CHAPTER 4

No Matter What, This Has to Stop!

Sometimes memories of traumatic events can be manipulated for political purposes, but sometimes immediate experience can trump memory. This is precisely what happened when newborn genetic screening (for phenylketonuria) and adult carrier screening (for Tay-Sachs disease) were first introduced. A visitor from Mars who knew something about the history of the Nazis might have expected both of these screening programs to be very controversial: the first because it was done under governmental mandate and violated the ethical norm of informed consent; the second because it was focused on the community that had suffered the worst abuses of Nazi eugenics, Ashkenazi Jews. In fact, neither program unleashed a storm of protest and, in the end, both became quite successful. Today, they are well-established social institutions, with their own sustaining procedures, funding patterns, and bureaucracies.

Screening programs search for biological needles hiding in populational haystacks. Tuberculin screening is a commonplace example: a non-invasive test is administered to many people in the same demographic group (children who are starting school, for example, or men and women being inducted into the armed services) in the hope of finding those very few individuals who are, in some way, afflicted with a worrisome condition. Newborn screening was instituted to test large numbers of babies in order to find the very few who suffer with phenylketonuria (PKU), an inborn error of metabolism which does not manifest itself clearly at birth. Carrier screening for Tay-Sachs disease was instituted to test large numbers of Jewish adolescents and adults in order to find the rare person who, without knowing it, carries a single copy of the gene that, when it is doubled, causes a fatal form of neurological degeneration. Like prenatal diagnosis, newborn screening and carrier screening emerged before the Human Genome Project began. These early tests did not, indeed could not, search for genes, because the genes that cause the diseases in question had not yet been located. Instead, each was a test for a unique chemical which, either by its presence or by its absence, indicated a disease state or a potential disease state, much the same way that a low level of iron in the blood signifies anemia or a high level of cholesterol indicates a high possibility of heart disease. Unlike prenatal diagnosis, however, neither newborn screening for PKU nor carrier screening for Tay-Sachs disease has raised the ire of feminists, left-wing intellectuals, patient advocates, or even members of the target populations—although both diseases are transmitted genetically and both produce mental and neurological degeneration.

The Story of PKU

The PKU story begins in Norway, in 1933—the same year that the Nazis came to power just across the Baltic Sea. A desperate Norwegian mother paid a visit to an experienced biochemist, Asbjørn Følling. The biochemist was exceptionally well trained; he had studied medicine at the University of Oslo and had then spent several postdoctoral years in the United States and Germany learning and practicing clinical biochemistry. In 1933 he was running the clinical laboratories of the Veterinary School at the University of Oslo.

The woman who visited him had two severely retarded children. Her husband, a dentist, believed there was an odd substance in his children's urine; the couple hoped that Følling could tell them what it was—and how it might be related to the retardation.[1] Using various chemical procedures, Følling soon discovered that both urine samples contained phenylpyruvic acid (which is not found in normal urine), and that it was easy to test for the presence of this compound. When a few drops of ferric chloride, a common laboratory reagent, were added, the urine immediately turned green.

Følling was fascinated. A research program had fallen right into his lap, and he pursued it for the rest of his working life. His initial efforts involved visiting various Norwegian institutions for the mentally handicapped and testing patient urine; out of 430 samples, ten tested positive for phenylpyruvic acid. In 1934 Følling published a paper announcing this discovery in a leading German biochemical journal. Several of the ten patients who had tested positive were siblings, which suggested to him that the disease they shared was probably hereditary. Følling also guessed that these patients were retarded because they did not completely metabolize phenylalanine (an amino acid found in many food proteins) and it was accumulating in their brains.[2]

One of the people who read Følling's article was an English physician, Lionel Penrose. Penrose had been raised in a devout Quaker household. Although he had become an atheist in his twenties, he remained deeply committed to the social values of pacifism and equality. His pacifism led him to serve in a Friends Ambulance Train Unit during World War I. His conviction that each person—no matter what his or her personality, intelligence, or status may be—has social value led him to be skeptical about eugenic doctrines of biological degeneracy. He thought, as one scholar has succinctly put it, "that the advocacy of sterilization revealed more about the neuroses of its proponents than about any behavioral tendencies among its objects."[3]

Penrose had an abiding interest in psychology and psychiatry, and in 1934 he was deeply engaged in studying various forms of mental deficiency. In 1930 he had been hired to do a careful study of the patients (numbering slightly more than a thousand) at the Royal Eastern Counties Institution in Colchester, England, a residential facility for the mentally deficient. His salary was being paid, in part, by the Medical Research Council of Great Britain. The stated goal of the project was to classify the different causes of mental deficiency; the unstated goal was to demonstrate that the eugenicists who argued that feeblemindedness was a Mendelian dominant trait were, to put it simply, wrong.

Penrose's task was to get to know each of the patients at Colchester and to classify the differences he found among them, but within two years he realized that this was not going to be easy. None of the standard measurements of intelligence seemed entirely satisfactory. Even if they had been, they provided no insight into the origins of the person's condition. Origins were exceedingly difficult to discern, partly because it was very hard to tell whether a person had been mentally deficient from birth (some parents could not be located, others were not forthcoming with information, while others had, understandably, no idea how to gauge the mental acuity of an infant or whether the advent of a particular disease or the occurrence of a particular accident had predated their child's difficulties).

Thus, when Penrose read Følling's article, he was immediately intrigued, and he set out to test the urine of the Colchester patients. Out of the thousand or so patients, he found two whose urine turned green when tested. The family histories of these two patients strongly suggested to Penrose that the condition was, as Følling had guessed, a rare but nonetheless Mendelian recessive; an inborn error of metabolism.[4] Working with biochemists, Penrose was able to push the boundaries of knowledge about the condition a little further. Postmortem examination of the livers of people with this condition revealed the lack of a crucial enzyme which normally converts phenylalanine to tyrosine, another amino acid. Penrose guessed that the problem in the patients' livers created the problem in the patients' brains, either because a substance that the developing brain needed was not produced or because the phenylpyruvic acid that wound up in the blood (and then in the urine) somehow damaged the developing brain.

Penrose recognized that PKU, as the disease had quickly come to be called, was not a major cause of mental deficiency, but he was excited about the finding nonetheless. Having spent the previous four years working in a mental institution, he knew all too well that every parent, physician, nurse, or social worker with whom he spoke was fatalistic: no one knew what caused retardation, and no one knew what, beyond custodial care, to do about it. If PKU was a metabolic condition, there might be a metabolic therapy: the removal of phenylalanine from the patients' diet. The possibility that PKU could be cured was a ray of hope in an otherwise dismal landscape. At Colchester, Penrose reported to the director of the Royal Eastern Counties Institution, Frank Douglas Turner, a physician whose son was mentally impaired. "I remember," Penrose wrote years later, "how Dr. Turner's eyes lighted up with excitement at this news and we went on to discuss the possibility in the future of rational treatment for such patients by altering their diet at an early age."[5]

Exciting prospects, yes, but they did not bear fruit immediately. Creating a phenylalanine-free diet was much easier said than done, and—perhaps because the condition was so very rare or perhaps because World War II diverted researchers' attention—no one made the necessary effort for almost twenty years. The person who finally did, a Swiss-born pediatrician, was pushed into taking action by another distraught mother.

In 1949 Horst Bickel was a junior assistant in the University

Children's Hospital in Zurich, Switzerland. His supervisor had instructed him to test the urine of every retarded child admitted to the hospital. When he moved, later that year, to a research fellowship in pediatrics in Birmingham, England, Bickel suggested to his new colleagues that they do the same thing. The third child he tested, a two-year-old, tested positive for PKU. "Her mother was not at all impressed," Bickel reminisced, "when I showed her proudly my beautiful paper chromatogram with the very strong phenylalanine (Phe) spot in the urine of her daughter proving the diagnosis. She awaited me every morning in front of the laboratory asking me impatiently when I would at last find a way to help . . . She was very upset and did not accept the fact that at the time no treatment was known for PKU. Couldn't I find one?"[6]

Bickel and his associates succeeded in creating a phenylalanine-free foodstuff by repeatedly heating a protein by-product (casein hydrosylate) with activated, acid-washed charcoal. They fed this unpalatable material to the child for the next several months, and her condition improved markedly. When they took her off the food she deteriorated quickly, "and within a few days she reverted to her original state." In the next several years progress was made: a food-processing company, Mead Johnson, was convinced to commercially produce a low-phenylalanine diet powder, and the urine of retarded children was regularly screened, in dozens of hospitals and residential facilities in several Western countries, to discern if any of them had PKU and might benefit from being placed on this special diet.

Newborns were also being tested, and therein lies the genesis of population screening for PKU. Several pediatricians who knew about PKU and the possibility of a dietary therapy began using the "wet diaper test"—putting ferric chloride on a baby's diaper—on all the newborns in their practice, in hope of identifying infants who would benefit from starting the low-phenylalanine diet before the disease could do significant neurological damage. Although the

Children's Bureau of the United States Department of Health encouraged and supported some of those efforts, everyone involved soon realized that when used on newborns the ferric chloride test was inadequate: there were far too many false positives. Their hope of diminishing, however slightly, the numbers of retarded children was—at least for the moment—dashed.[7]

Thus for almost three decades very few people were helped by Følling's original discovery. Between 1935 and 1955 a few parents of PKU children may have been warned off having additional children; between 1955 and 1960 a few hundred retarded children may have been less severely damaged by the disease because they were placed on a low-phenylalanine (low-Phe) diet. A genetic screening program for PKU did not begin until the early 1960s, when a more reliable diagnostic test was developed for newborns.

Robert Guthrie was a physician involved in cancer research when, in the mid-1950s, he and his wife were told that their second child was mentally retarded. Not surprisingly, Guthrie soon developed an interest in the clinical care of what would now be called children with developmental disabilities. He found a job at the Children's Hospital in Buffalo, New York. One of the pediatricians there, who was experimenting with low-Phe diets, told Guthrie that the diet was most effective when levels of Phe in the child's blood were monitored, but that there was no easy way to do the monitoring.

Guthrie had an idea about how a simple blood test could be developed. He knew that a type of test, bacterial inhibition assay, was used to screen for various substances in the blood of patients undergoing cancer treatment, and he thought he might use the same principle: "a compound which normally prevents growth of bacteria in culture plates," as he explained, years later, "no longer inhibits the growth when large amounts of Phe are present in a blood spot that is added to the plate."[8]

While Guthrie was working to perfect this test, one of his nieces was diagnosed with PKU at the age of fifteen months. "She was developmentally delayed. A physician tested her urine with ferric chloride . . . Her urine tested positive, but unfortunately it was too late to prevent the mental retardation." Guthrie realized that the test he was developing could have provided the diagnosis earlier in his niece's life, because a high concentration of Phe would appear in a newborn's blood before it would appear in her urine. He also knew that "the earlier dietary treatment began the better the prognosis for intellectual growth." He tested the blood of residents of a nearby school for the mentally retarded. "I identified all who were already known from urine testing to have PKU, and four additional patients who had not been detected previously by urine screening." Guthrie was excited by a somewhat different prospect from the one that had inspired Lionel Penrose thirty years earlier: with a test that could detect PKU within a few days of birth and a specially prepared diet administered early on, the mental retardation of PKU might be wholly averted, not merely remediated.

What later came to be called the "Guthrie test" was reasonably simple and direct; the equipment needed already existed in most clinical laboratories, which meant that technicians would not even need special training; and the dietary powder was already being produced commercially. Pediatricians thought it would be a simple matter to get a few drops of blood from the heels of babies before they were sent home from the hospital. All that was needed was the political will to make a population screening program happen—and Guthrie was already affiliated with an organization that had not just the political will but also the political skill for the job: the National Association of Parents and Friends of Mentally Retarded Children (later the National Association for Retarded Citizens, NARC, and now The ARC).

In 1961, when Guthrie had sufficiently refined his assay, NARC was already a decade old. It had started as a loose confederation of fifty-seven local organizations founded by parents of children who were mentally deficient. Most of these parents were well-educated, middle-class people. They had first met one another—and decided to band together—because they had been attending meetings of the American Association on Mental Deficiency (AAMD), a professional association of physicians, psychiatrists, and social workers. AAMD was, they felt, all well and good, but the questions that clinicians were asking were not the questions to which parents wanted answers. These parents felt "an acute lack of community services for retarded persons." They were unhappy about "long waiting lists for admission to residential institutions," and also about the fact that public school systems excluded "children with IQs below 50."[9] Following the model pioneered by the March of Dimes and the American Cancer Society, the parents who organized NARC believed that by banding together they could "bring major benefits in public relations, exchange of information and political action"—as long as they had the assistance of a few key professionals.

They turned out to be right. In 1952 NARC created a research advisory board, headed by a pediatrician from Yale Medical School; two years later the organization had enough money to hire an executive director and to open a national headquarters in New York. That same year it also managed to get the Eisenhower administration to proclaim the first National Retarded Children's Week. By 1955 NARC had 29,000 individual members and 412 local units. Some of those units had become so active that their work could no longer be carried out solely by volunteers: "28 units reported that they employed an executive director or secretary either full of part-time." More than 200 units had started classes for retarded children who were educable; 86 had recreational or social groups for the children; 78 had counseling and guidance services for parents. In order to be more effective politically, 138 of the units had legislative committees and 163 had become affiliated with local councils of health and welfare agencies.

By 1961 NARC had tested its political and publicity muscles several times, with local, state, and federal agencies and in state and federal legislatures. It had produced several educational films, run newspaper advertisements, and learned how to place articles in magazines. It had also published several "action agendas" and research reports and was maintaining a regular newsletter for communicating with members. It had worked with celebrity parents of retarded children—Pearl S. Buck, Roy Rogers, and Dale Evans—to publicize their books and to make clear the fact that mental retardation is not just a disease of the "poorer classes." Members of an affiliated organization (AHRC) had held a virtual sit-down strike in a state capitol; when the governor of New York refused to see a delegation of parents and children, the parents told reporters they would leave their children in the outer office so that the governor's staff could take care of them for a day.[10] NARC had several friends in Washington, both in Congress and in the Children's Bureau. After January 1961 it also had a friend in the White House; John F. Kennedy's sister Rosemary Kennedy was retarded, and his family was active in creating and funding programs for retarded individuals. NARC had argued in favor of social security coverage for adults disabled in childhood and additional federal funding for medical facilities and vocational rehabilitation. Members had testified in Congress and sent delegates to White House conferences. NARC's membership had mushroomed to 62,000, and it was one of only ten voluntary health organizations permitted, by executive order, to raise funds from federal employees.

NARC had also funded Guthrie's development of a bacterial inhibition assay. When Guthrie was satisfied that he had finally developed a sensitive assay, the president and the executive director of NARC were the first to know. After that, things moved very fast. NARC urged Guthrie to publish his results as quickly as possible—as a letter to the editor of a medical journal, not as a research report—so that it could be publicized as part of NARC's 1961 poster campaign.[11] The poster depicted a pair of sisters, and it carried this text: "Sheila is retarded. Kammy was spared because of new medical research. Retarded Children *Can* Be Helped."[12]

Within the year, the Children's Bureau of the United States Department of Labor, which had a long-standing relationship with NARC and a long-standing interest in alleviating the problems associated with mental retardation, had decided to fund a national trial of newborn screening. Using those funds, Guthrie hired a staff and began preparing PKU test kits for examining one million infants. In two years' time, four hundred thousand infants were tested and thirty-nine cases of PKU were identified—and retesting revealed that none had been missed. "It soon became apparent," Guthrie later wrote, "that the old urine screening method was less effective than blood Phe screening."[13]

Guthrie quickly became very active in promoting the idea of mass newborn screening. "My father had been a traveling salesman and I must have inherited his genes, because I have always felt challenged by what he would have called the 'hard sell.' I accepted every opportunity to travel and speak about the need for the screening test to detect *and treat* newborn infants with PKU."[14] With funds provided by the Presidential Commission on Mental Retardation and the Joseph P. Kennedy Foundation, the Advertising Council (a consortium of American advertising agencies that provides creative services to philanthropic organizations) had mounted a campaign to advocate for population screening of newborns as "a must for all babies everywhere," urging parents to demand that such screening be publicly funded.

NARC also went into high gear; the central office advised local units to begin lobbying state legislatures for laws mandating PKU screening. By the time a peer-reviewed article about the assay finally appeared in 1963, three states had already passed such legislation (Massachusetts, Rhode Island, and New York) and the Presidential Advisory Commission on Mental Retardation had already suggested that newborn screening programs be created.[15] In Massa-

A fifty cent test spared this baby from spending his life mentally retarded.

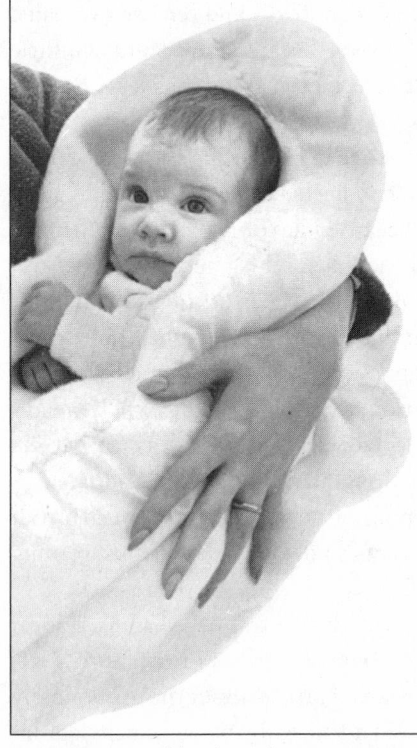

But only in Massachusetts, New York, Louisiana and Rhode Island is the test legally required.

There is a disorder of body chemistry called phenylketonuria, or PKU. Once, there was no way to diagnose it. Once, babies born with it were doomed to lifelong mental retardation.

But no longer. For a new test—given during the first few days of life—detects PKU *before* brain damage occurs.

A special diet corrects the trouble. A PKU baby then grows and develops as normally as any other child.

This test costs only 50¢ a baby. But a PKU victim, undiscovered, can cost $100,000 for lifetime care in an institution.

If your state doesn't require the PKU test, urge its adoption. It should be a must for all babies everywhere.

Here are six things you can do now to help prevent mental retardation and bring new hope to the 5½ million people whose minds are retarded:

1. If you expect a baby, stay under a doctor's or a hospital's care. Urge all expectant mothers to do so. Take your baby to the doctor regularly for health checkups.

2. Visit local schools and urge them to provide special teachers and special classes to identify and help mentally retarded children early in their lives.

3. Urge your community to set up workshops to train retardates who are capable of employment.

4. Select jobs in your company that the mentally retarded can fill, and hire them.

5. Accept the mentally retarded as American citizens. Give them a chance to live useful, dignified lives in your community.

6. Write for the free booklet to the President's Committee on Mental Retardation, Washington, D.C.

A poster created by the Advertising Council advocating mandatory screening for PKU, as well as nondiscriminatory treatment for retarded children and adults.

Courtesy of Ad Council Archives, University of Illinois Archives, RS 13/2/207.

chusetts, the first state to mandate testing, a voluntary program, involving every maternity hospital in the state and funded by the state department of public health, had already been functioning for a year before the legislation was passed. By 1965 thirty additional states had signed on—even before the results of the initial trials, sponsored by the Children's Bureau, had been published.

Today virtually all states mandate newborn screening, not just for PKU but for several other conditions, most of which are inborn errors of metabolism even rarer than PKU. Newborn genetic screening is also standard practice in Canada, Britain, most of Western Europe and Scandinavia, much of Central and Eastern Europe, Israel, Egypt, Japan, Australia, and New Zealand—almost all of what have come to be called the developed countries of the world.

Following his success with PKU, Guthrie and his colleagues developed bacterial inhibition assays for several other inborn errors of metabolism (for example, galactosemia and homocystinuria), and these tests also became a part of standard pediatric practice; recently bacterial inhibition assays have been supplanted by an even more sensitive (and less expensive) technological system: analysis by tandem mass spectrometry. Since the early 1960s millions upon millions of babies have had a drop of blood taken from a heel a day or two after birth so that the presence of an inborn error of metabolism could be detected and—presumably—so that effective therapy could be undertaken or the parents alerted not only about disease symptoms but also about their own status as carriers.

Newborn screening programs were not without their critics, and they remain mildly controversial, albeit only among professionals. In the 1960s many state medical societies objected to mandated screening because they felt that governments should not legislate medical policy except when a real public health crisis existed—and a very rare disease did not, they thought, bear any resemblance to a public health crisis. Medical researchers actively involved in studying metabolic diseases objected because the evidence that dietary therapies would prevent mental retardation was skimpy at best. No controlled trials of the diet had been done, and no one knew when or if the diet could be stopped—or even how long a child could remain healthy on such a bizarre nutritional regimen. Worse yet, the Guthrie test, although very accurate overall, occasionally produced some false negatives and false positives—which meant, these researchers feared, that some PKU infants were missed and also that some children were unnecessarily subjected to therapeutic diets that might harm or kill them. Researchers also worried, rightly, that with mass screening and widespread use of the dietary regimen it would become impossible to discover what it was about the inability to metabolize phenylalanine that damaged the developing brain—which meant that screening could prevent a full, medically rational understanding of the disease.[16] In short, some members of the scientific and medical communities initially objected to newborn screening for these genetic diseases, whether mandated or voluntary, as policies based more on emotion than on information, more on clever public relations than on science, more on technical capacity than on scientific rigor.

In the early 1970s, after some—but by no means all—of these initial fears had been assuaged, new ones arose. When newborn screening programs were first put into place, no one seemed to care about the fact that diagnostic tests would be performed on infants without their parents' consent. In those days, diagnostic tests and all kinds of therapies were regularly administered to patients, of all ages, without their consent and sometimes without their knowledge. Public discussion of the need for informed consent intensified, first in the late 1960s (with regard to undisclosed side effects of the birth control pill) and then in the early 1970s (when the Tuskegee syphilis experiment and other scandals involving human beings as research subjects—particularly institutionalized and mentally deficient human beings—were revealed). A few politically liberal medical geneticists who were involved in newborn screening were among the first to call attention to the widespread failure to inform parents and ask for their consent to the procedure; they advocated that the mandating laws be amended to require parental consent. Maryland instituted such a provision in 1976, but only one other state (Wyoming) and no other nations have followed suit. Many medical geneticists have argued that the heel-prick test itself is essentially harmless, that informed consent would be difficult to obtain from the parents in the confusion of the first days of their babies' lives, that a large number of refusals would vitiate the public health rationale for testing, and—ultimately—that the risk of harm to the child who does not get therapy is greater than the risk of harm to the autonomy of the parents whose child is tested without their consent.[17] Informed consent is much more of a concern in the United States than in any other country, but even in the United States consent for newborn screening is not a hot-button issue, and most ethicists and legislators have sided—when they have bothered to consider the issue—with the majority of professionals.[18]

Screening of newborns for genetic disease remains, to this day, a matter of some concern to bioethicists, to some clinicians, and to students of biomedical policy. Some people wonder whether it is right for public funds to pay for screening but not (in the United States at least) for the treatment, which is the sole justification for the screening. Others worry about the ethical rationale of spending millions of dollars every year to identify a very few potentially disabled infants, when the health needs of millions of other infants go unmet. Some medical geneticists are troubled when new technologies make it possible to screen for diseases for which no treatment exists; others are unhappy about the potential invasions of privacy that could occur when blood samples are saved in data banks and identifying labels (of necessity in a screening program) become attached to the samples.

People who know the details of the PKU story become disturbed when biology textbooks refer to PKU screening as a "complete success," since it was not that when it began and still is not today. The diet alleviates but does not eliminate neurological damage, and many individuals and families find it extremely difficult to comply with its demands; in the United States many insurance companies still do not cover the cost of the diet completely. Also, many medical geneticists know that while it is easy to test large numbers of babies in the hospital, it is far harder to track the afflicted babies after they have gone home—and even harder to convince some parents that expensive and strange therapies are imperative for their normal-looking children. On top of all this, when the first generation of young women with PKU who had been sustained on the diet began to have children of their own, their babies were born severely brain damaged unless they had adhered to their diet scrupulously during pregnancy.

These problems have not, however, led to a public outcry, and certainly not to a widespread movement to end neonatal screening—which expands, in terms of the number of babies tested and the number of conditions tested for, with every passing year. Numerous criticisms have been leveled by knowledgeable critics who come from many different countries, ethnicities, religions, and political persuasions, but none of them has ever claimed that any of these programs was eugenic and no one has ever suggested (in print, at least) that the reason to test a newborn for PKU—or any other inborn error of metabolism—was to sterilize (or otherwise end the reproductive lives of) its parents.

Quite the contrary: every single one of the critics of neonatal screening has acknowledged that the advocates of screening have acted out of ethical motives, the principal one being a desire to alleviate suffering and, in the case of diseases that cause damage to the brain or the nervous system, to maximize human potential. The initial newborn screening program on which all others were modeled was the one for PKU. The legislation that led to the founding and funding of these programs was often written by and always lobbied for by the parents and relatives of mentally compromised children, people like the founders of NARC, who worked long and hard to alleviate the social and personal problems caused by mental retardation. Prevention of mental retardation through newborn screening and dietary therapy seemed to them to be a wholly admirable and ethically justified goal. No one ever suggested that the programs should be shut down because parents and healthy siblings were being stigmatized or because preventing this form of mental retardation would, somehow, increase discrimination against the mentally disabled. In the case of newborn screening, the experience of caring for mentally compromised people was more than sufficient to outweigh the memory of what the Nazis had once done in the name of eugenics.

The Story of Tay-Sachs Disease

Tay-Sachs disease is named after the two physicians who first identified its clinical signs. Warren Tay was a British ophthalmologist. Three partially blind infants were brought to a clinic he supervised in London in the early 1880s. Each child had the same combination of symptoms: severe muscular weakness (none of the children was able to turn over or to sit up) and an inability to respond to objects and people (although each was able to follow a moving light). Using a newly invented instrument, the ophthalmoscope, Tay looked at the babies' retinas and was startled to see something he had never observed before: a bright red spot on the fovea, the center of the retina, just in front of the end of the optic nerve. The first two children died within a year after Tay examined them. A few years later the parents of one of the children brought another infant with exactly the same symptoms to Tay's clinic; Tay began to think that the disorder, whatever it was, might be hereditary.[19]

In New York, at just about the same time, Bernard Sachs, a young physician who had just begun to practice, examined an infant with very similar symptoms: poor muscular development, poor eyesight, unresponsive to normal stimuli. Sachs was the son of German-Jewish immigrant parents; he was fluent in German and had decided, after graduating from Harvard College, to go to Germany for his medical training. In those years (the late 1870s) German physicians were the world's foremost advocates of pathological anatomy, which involved postmortem examination of tissue specimens under the microscope. When Sachs returned to New York in 1884 he became a junior partner in the practice of another German-Jewish physician. The senior partner, knowing that the younger man was very interested in neurology, passed this patient to him. The child, a girl, appeared to be normal at birth, "its features beautifully regular." But by the time she was about two months old her parents had become worried. "The child was much more listless than children of that age are apt to be . . . it took no notice of anything." An ophthalmologist who examined the child at that time saw a cherry red spot on her fovea, but since there was no damage to any of the retinal tissue, he predicted that she would soon develop normally. That was not to be. When Sachs examined the baby, more than a year later, he reported that the child had "never attempted any voluntary movements and . . . as it grew older, gave no signs of increasing mental vigor . . . It could not be made to play with any toy, did not recognize people's voices and showed no preference for any person around it." By that time the baby girl was also completely blind.[20]

The young neurologist must have developed a trusting relationship with the parents, because when the child died—just before her second birthday—they sent the body to him for a postmortem examination. Sachs dissected the child's brain and, finding that the gross anatomy of the cerebrum was abnormal, removed some tissue from the cerebral cortex for analysis under the microscope. The cells he saw were profoundly abnormal. "In my search through the entire brain I have not come across more than a half dozen . . . cells of anything like normal appearance . . . The contours are rounded [normal cerebral cells have spiked projections on their edges] and the cell substance exhibits every possible change of its protoplasmic substance." In particular, Sachs noted, there were "detritus like" masses within the cell walls.[21]

Over the course of the next decade Sachs examined several other children with the same symptoms and the same postmortem results. He also read similar case reports in the medical literature, including Tay's. By 1896 he was ready to give this newly discovered disease entity a name: *amaurotic family idiocy: idiocy* because the child never developed normal mental functions, *amaurotic* because what started out as poor vision eventually became complete blindness, and *familial* because all the cases reported in the literature were—to use Sachs's term—"Hebrews." Since several of the afflicted children were siblings and a fair number of the parents were in consanguineous marriages (the parents were first cousins), Sachs concluded that the disease was probably hereditary, or, as he put it, "heredodegenerative."[22] It was also, invariably, fatal.

In the following decades researchers learned a bit more about amaurotic family idiocy. Most reported cases were children of Jews from Eastern Europe, Ashkenazi Jews, but by 1910 physicians had reported a few non-Jewish cases, particularly in northern New England and Quebec, among people of French-Canadian stock. By that time the stains used on microscopic specimens had become more sophisticated, and several clinicians, Sachs among them, had begun to suspect that the "detritus" that destroyed the cortical cells of these doomed babies was a fatty material, a lipid of some sort. There was also some debate in the medical literature about whether the disease was hereditary. Some physicians thought it might be the result of a toxic agent absorbed by the infant.[23]

In 1933, however, a researcher from the Galton Laboratory, David Slome, used a form of biostatistical analysis on 135 reported cases to demonstrate that the disease was most likely a "single recessive gene substitution."[24] A few years after that a German biochemist, Ernst Klenk, working with a large collection of postmortem tissue samples, discovered that the lipid that accumulated in the brains of babies with Tay-Sachs disease was actually a ganglioside—a complex molecule consisting partly of a fat, partly of a sugar, and partly of a protein derivative.[25]

So matters stood until the postwar years, when a new generation of clinical researchers in the United States, well versed in biochemistry and in biochemical laboratory techniques, began to push at the boundaries of pathological knowledge. These researchers had novel forms of assistance: a government that was newly willing to fund basic medical research, hospital administrators who were newly willing to allocate beds (even when the patients could not pay) for research projects, and, crucially, an organization of medical activists—the National Tay-Sachs Disease Association (NTSDA)—eager to sponsor research that might lead to a cure.

NTSDA was founded in 1957 by a group of parents of babies who had been diagnosed with amaurotic family idiocy (a name they refused to use; hence the new name, Tay-Sachs disease); they had met through the physicians who were caring for their children. The physicians and the parents were all Jewish. Some had been victims of the Nazis, having fled Eastern Europe in the 1930s or having had family members killed in the camps. Others had been in the armed forces, fighting the Nazis, during World War II. They belonged not only to an ethnic community but to a generational community as well. As parents of young children, they had witnessed the beginnings of the successful fight against polio (mass inoculation of schoolchildren with polio vaccine had started in 1954). They had also witnessed the demise of epidemic tuberculosis that came with the widespread civilian use of antibiotics after 1945. They must have known how crucial various medical philanthropies had been to those medical triumphs; how the March of Dimes and the American Lung Association had galvanized public attention and public monies. And, as Jews, they must have known how easy it had become to raise money from their peers for the support of Jewish causes.

And so, in a spirit of both mourning and redemption, the parents banded together to defeat Tay-Sachs. "I wanted to feel," as one parent later put it, "that my baby did not die in vain—and that no other family would ever have to suffer the way we did."[26] In the mid-1950s they created local organizations and then, in 1957, a national federation. They appointed the physicians who had cared for their children to the medical advisory board. They created a newsletter; they sold raffle tickets; they went door to door; they spoke at synagogue services; they organized dinner dances; they invited celebrities; they held auctions—copying techniques that had been pioneered by other medical and Jewish philanthropies. They raised as much money as they could, and gave it, in the form of small research grants, to the physicians. Wise to the ways of research, they also tried to leverage their money by locating basic researchers who were already funded by NIH, whose laboratories could be asked to focus on Tay-Sachs. Periodically, they helped organize symposia, so that the researchers could talk with one another—and with parents. The parents also volunteered themselves and their children—both those who were afflicted and those who were not—as research subjects, freely donating blood samples, tissue samples, genealogies—whatever it took to find the underlying disease mechanism so as to find a cure. Echoing national politics of the 1960s, their motto was "We will reach for the moon."[27]

In a little more than a decade, all this concerted effort paid off. A cure proved elusive, but the activists found something they regarded as almost as good: a relatively easy way to locate carriers of the defective gene.

In 1962 biochemical researchers had identified all the component parts of the ganglioside that accumulates in the cerebral neurons of Tay-Sachs children.[28] The next step was a bit more difficult: determining whether the disease was a result of overproduction of the ganglioside or failure to degrade it during cellular metabolism. Biochemical studies of the breakdown products, combined with comparisons between the blood of the parents and that of their normal and afflicted children, soon led biochemists to conclude that the root of the problem was a missing enzyme that, if it were present, would degrade the Tay-Sachs ganglioside. In 1969 a team of biochemists in California definitively identified the missing enzyme: *β*-*N*-acetylhexosaminidase, now simply referred to as Hex-A.[29] When Hex-A is present and active in a cell, it breaks down the Tay-Sachs ganglioside by cleaving the sugars in the compound—a normal part of cellular metabolism. Parents and some siblings of Tay-Sachs babies had levels of Hex-A that were below normal, but afflicted children had no Hex-A at all. Without it, the ganglioside accumulated in their tissues and, as it piled up, it compromised their brains.

Unfortunately, no matter how they tried to administer Hex-A to afflicted babies—whether by injection or by mouth or by serum transfusion—the accumulated ganglioside did not disappear, nor did the buildup of the chemical subside. All kinds of things turned out to be wrong about Hex-A as a cure: it was a large molecule that did not easily cross the boundary between blood vessels and body tissues; for the same reason, it could not be given to pregnant women in the hope of having a positive effect on their fetuses. Hex-A, it turned out, was also digested very fast; within twenty-four hours after it was administered to babies, no traces of it could be found in their bodies. Hex-A deficiency was clearly the cause of Tay-Sachs disease, but, just as clearly, Hex-A supplementation was not going to work as a cure.

Nevertheless, as one of the founders of NTSDA put it, if Hex-A couldn't be made to work as a cure, it could be made to work "as a control."[30] They referred, in part, to prenatal diagnosis; if you could apply an assay for Hex-A to fetal cells, you could tell parents if their child would be afflicted with Tay-Sachs, and then they could decide whether or not to terminate the pregnancy. Better still, a test for Hex-A would tell the siblings of an afflicted baby whether they were carriers (on average, two-thirds of siblings would be; one-third would not) and thus at risk for bearing their own afflicted child. In short, "control" meant that with a test for Hex-A, Tay-Sachs disease could be prevented—either by premarital screening and reproductive caution or by prenatal diagnosis and abortion. This was not the moon that the founders of NTSDA were reaching for, but it was progress nonetheless.

Shortly thereafter, two reports of successful amniocenteses for Tay-Sachs fetuses appeared in the literature. The authors of one report were colleagues of two of the early Tay-Sachs researchers at Downstate Medical School in Brooklyn, New York. Two authors of the other report were the biochemists John S. O'Brien and Shintaro Okada, who had discovered Hex-A deficiency.[31] Each woman who subjected herself to the amniotic tap had previously given birth to a Tay-Sachs baby. The diagnoses were deemed successful, because when no Hex-A was found in the fetal cells and the pregnancy was terminated, postmortem examination of the fetus revealed abnormally high levels of the Tay-Sachs ganglioside in many tissues. Ruth Dunkell, founding president of NTSDA, later recalled that she and her husband "were thrilled, not because we wanted more children—we were done—but because we knew so many people who did."[32] Indeed, as soon as the Hex-A test was perfected, many members of the NTSDA had the carrier status of their healthy children determined. Their reason was straightforward: to spare their children the agony of bearing a Tay-Sachs child. If the child was found to be a carrier, they reasoned, the child's eventual partner could be tested also; two carriers would be able to use prenatal diagnosis and abortion to ensure that the family tragedy was not repeated. This was not a perfect solution to the problem of Tay-Sachs disease, but it was better, in their view, than having to care for and then mourn a suffering baby.

Like the men and women who founded the NTSDA, the clinician-researcher who managed to turn individual testing of possible carriers into a community-wide screening program had experienced an emotionally devastating encounter with the disease. In 1969, not long after he joined the faculty of the Johns Hopkins University Medical School as a medical geneticist, Michael Kaback got a call from a pediatric neurologist who had just examined a colleague's ten-month-old son. The child had been regressing, and all clinical signs pointed to Tay-Sachs. The neurologist asked Kaback if he knew a way to confirm the diagnosis. Kaback knew about the Hex-A test because John O'Brien, one of the biochemists who had discovered it, had recently visited Hopkins to run the test on several of Kaback's Tay-Sachs patients. The afflicted child's blood was sent to O'Brien's lab for analysis and in a few days the diagnosis was confirmed.

"Perhaps the most powerful aspect of that diagnosis was that Karen, the infant's mother, was 7 months pregnant with their second child at that time," Kaback later recalled. "The anxiety and sadness that prevailed were beyond description."[33] The pregnancy was too far advanced for the couple to consider abortion. They were devastated by the notion that their next child would deteriorate and die just as their first would. So, together, the physicians and the couple made a heartrending decision: the baby would be tested right after birth. "If the child did have TSD, then it would be placed in foster care . . . without the parents ever seeing it."

> In May, 1970, Karen delivered a full-term female infant. Her husband [Bob] and I were both with her in the delivery room . . . The scene in the delivery room . . . is one I shall never forget. Two wonderful young people covering their eyes, afraid to look at a beautiful crying baby girl, for fear that they would never see her again.

Some of the baby's cord blood remained in Kaback's lab for analysis; the rest was shipped by air to John O'Brien's lab in San Diego.

> At midnight that evening, I heard some movement in the hallway, outside my laboratory. It was [Bob] our young intern, the baby's father . . . he was not going to leave. We chatted as I worked . . . By 4:00 A.M., my results were in and they were unequivocal . . . A call to San Diego (about 1:00 A.M. there) reached John. He had just completed his . . . studies and his results were the same—an unaffected infant. At about 5:00 A.M., Bob and I went to see Karen. She was awake, unable to sleep. Together, the three of us walked to the pediatric floor, and they held their new baby for the first time. The joy, relief, and sheer ecstasy for this couple was of a magnitude like nothing I had ever witnessed.

Within hours, as Kaback recalled, "the impact of this experience really began to sink in. What had happened to this young family need never be replicated."

The assay for Hex-A was not easy to perform, but Kaback believed it could be at least partially automated—and he was sure that research funds could be found to accomplish that step. Tissue samples would be easy to obtain; all that was needed was a vial of blood from each person to be screened. It would not be easy, he thought, to find large numbers of adult volunteers, but it could be done—and the end result would more than repay the effort. At-risk couples could be identified and counseled and then, if they were willing (and Kaback, who is Jewish, knew his own peer group fairly well and was pretty sure that they would be willing), their pregnancies could be monitored through amniocentesis. If a large-scale carrier testing program could be established, Karen and Bob's ordeal would not have to be repeated. If the program worked, the overall number of Tay-Sachs births could be dramatically reduced, perhaps even to zero, at least among Ashkenazi Jews.

Almost a year passed while the various elements for mass screening—human and technological—were assembled. Serum from large numbers of Tay-Sachs parents (the "obligate heterozygotes" who were certain to be carriers) was subjected to the assay, so that Hex-A levels among carriers could be determined. The assay itself was semi-automated, so that large numbers of samples could be tested efficiently and accurately. Rabbis in the Baltimore-Washington corridor were contacted and given information about Tay-Sachs disease and about the proposed testing program. The rabbis were asked to provide space for a day of drawing blood samples; they were also asked to urge their congregants to be tested, particularly those who had not yet had children or who had not yet completed their families. Groups of volunteers from Jewish charitable organizations were recruited and trained to assist with the testing. Funds were raised, with the help of the National Capital Tay-Sachs Foundation, from private sources in the community, from an endowment that the Kennedy family had created at Johns Hopkins, and from the Maryland State Department of Health and Mental Hygiene. Informational stories were placed in the local media, including newspapers, radio, and television. Local physicians and phlebotomists were recruited to take the samples and local geneticists and social workers to provide counseling.

On a spring morning in 1971 the first community-based Tay-Sachs screening program was launched in the basement of a Reform synagogue, Temple Beth-El, in Baltimore. Eighteen hundred people presented themselves for screening during a seven-hour period. All were asked, and most agreed, to pay a five-dollar service fee. Unmarried individuals were counseled before the samples were taken—and most decided to wait until marriage to be tested. Two hundred and fifty pregnant women came, but only their husbands were tested (the Hex-A screen was not accurate during pregnancy). Of the 1,359 people who had their blood drawn, 98 percent were Jewish (the others were non-Jewish spouses of Jews). Thirty-eight of those people turned out to be carriers. That initial program was repeated, using different Jewish venues in the Baltimore-Washington area, eight times in the course of the next year. Altogether about seven thousand individuals were tested, and two hundred fifty individual carriers and eleven at-risk couples who had not yet borne a Tay-Sachs infant were identified.[34]

Kaback and his colleagues, pleased with what they had accomplished, quickly began publicizing their methods and their results. Within two years similar testing programs were introduced in other cities with large Jewish populations: Montreal, Toronto, Philadelphia, New York, and Miami. In 1973 synagogues in California successfully lobbied for the creation of a publicly funded, statewide program. By the year 2000, one hundred cities in fifteen countries on six continents had established testing programs for Tay-Sachs (and other, similar, single-gene recessive storage diseases). There is now an international quality assurance service, funded by the National Tay-Sachs Disease and Allied Disorders Association, which tracks all the community screening programs and conducts annual quality control assessments. In 1970 there were estimated to be between 50 and 60 Tay-Sachs births a year in the United States, 40 to 45 of them to couples of Jewish ancestry. By 1999, as a result of carrier and prenatal testing, the average total number of new Jewish cases a year was down to four: a 90 percent reduction.[35]

Many knowledgeable people regard these voluntary Tay-Sachs screening programs, based in Jewish institutions such as synagogues and community centers, not only as successful, but as exemplars of the way mass screening of adults ought to be conducted. Aside from the successful outcome, the crucial characteristic of these programs to which most commentators point is that fact that they were wholly community based. Most of the medical geneticists who organized them were Jewish; the places in which blood was collected were synagogues, schools, and community centers; many of the genetic counselors were Jewish; and the funding often came from Jewish philanthropies. In the case of Tay-Sachs, a devastating genetic disease was defeated in large part by the ethnic community that was plagued by it.

At the beginning, however, there were dissenters and difficulties. Within the Jewish communities themselves there were several kinds of objections. First, some people felt that Jews were already stigmatized and did not need to have an additional disease—particularly one that was genetic—attached to their cultural definition. Second, some Jewish leaders worried that unmarried carriers, once identified, would either decide not to have children at all or would decide to marry "out." In the post-Holocaust decades both these behaviors—intermarriage and suppressed fertility—were regarded by some leaders as inimical to the future of the Jewish community: "doing Hitler's job for him," as it were. These more or less ethnic concerns were mirrored, in the medical profession, by the belief that the communal benefit of preventing the birth of a few disabled babies was not worth the communal cost of that much psychological trauma and that much lost reproductive opportunity.[36] On top of this was a practical difficulty, which became apparent in each locality after the first flush of enthusiasm and success had waned: people forget health information quickly, especially if the information is in any way negative. In order to prevent Tay-Sachs birthrates from creeping back up in succeeding generations, community screening was going to have to continue in perpetuity—and that would be a very expensive proposition, not just in funds, but in volunteered time. In the end, none of these concerns proved to be a serious impediment to screening. Once informed, most Jewish young adults and most physicians voluntarily participated in these programs. Virtually all at-risk couples decided to undergo prenatal diagnosis and, when the fetus was afflicted, to terminate the pregnancy.

There was, however, one significant set of Jewish communities that objected to the screening programs vociferously. Ultra-Orthodox Jews are skeptical of medical experts who are not members of their own communities—and they are completely opposed both to abortion and to contraception. The early community-based screening programs for Tay-Sachs were targeted, deliberately, at couples who were already married—which meant that the medical geneticists were offering at-risk couples the opportunity for prenatal diagnosis followed by abortion. As a result, traditional Orthodox rabbis and their congregations had refused to participate in screening programs. By the early 1980s, according to two experts, "it became apparent that the [traditional] Orthodox and Hasidic communities were generally not participating. The birth rate of babies with Tay-Sachs disease began to decline in the 1970s, but this was not true among the children of religious Jews."[37]

In 1983 Rabbi Josef Ekstein, a leader of one such Orthodox group, learned, for the fourth time, that his newborn child had Tay-Sachs disease. "When you get one shock, then another, then another, either you give up and die or you fight back," he later told an interviewer. "I thought: 'I don't want another person or another family to go through that. No matter what, this has to stop.'"[38]

A Wedding Gift For You

Mazel tov on your recent engagement! We know this is a busy time for both of you, but we're asking you to add one more item to your pre-wedding checklist.

Please find the time to have your Tay-Sachs carrier test and to ask about Canavan screening. It only takes a few minutes to protect your future and your family.

National Tay-Sachs and allied Diseases Association of Delaware Valley encourages you to use the attached certifica te for two free Tay-Sachs and Canavan tests at your nearest testing site.

We wish you every happiness, long life and good health.

Gift of Life Certificate

Presented to

on ______ *date*

for two free Tay-Sachs and Canavan tests

To set up your appointment call:

Rabbi ______

Synagogue ______

Expiration date:
One Year from date of issue

The National Tay-Sachs and Allied Diseases Association of Delaware Valley promotes screening by asking rabbis to give couples this gift certificate during premarital counseling.

Courtesy of Rebecca Tantala for NTSAD-DV.

Working with the physician who had diagnosed his child's malady, Robert Desnick of the Mount Sinai School of Medicine in New York, Ekstein came up with an idea: test young people for carrier status while they are still in high school, before they even contemplate marriage, then create a confidential registry of assay results. When a marriage is arranged or contemplated, urge young people (or their parents, or the matchmaker) to call the registry to find out if the match is "genetically compatible." If both young people turn out to be carriers, provide genetic counseling, in the hope of convincing them not to proceed. Through such an arrangement, Tay-Sachs births could be avoided without recourse to either prenatal diagnosis or abortion.

Dor Yeshorim, the Committee for Prevention of Jewish Genetic Diseases, began offering educational, testing, and registry services in the New York area in the mid-1980s—and it has since spread to traditional communities in Canada, Europe, Australia, and Israel. Its services are entirely anonymous and confidential. Students who agree to be tested (about 90 percent of those who are enrolled in ultra-Orthodox parochial schools) are given a card with an identification number and a telephone number. These young people are not told their carrier status until they are contemplating marriage. When they call Dor Yeshorim, they are told whether their match is genetically compatible and, if it is not, they are offered genetic counseling. Rabbi Ekstein and his staff believe that "the vast majority of at-risk couples did not pursue marriage," and they base this judgment on "subsequent queries with different suitors."[39] As a result, Tay-Sachs disease has been almost entirely eliminated from these Orthodox communities around the world; the last baby treated in the special Tay-Sachs ward of the hospital that serves Brooklyn's Hasidic community died in 1996. The Dor Yeshorim program has had difficulty extending its reach into other Jewish communities, even Orthodox ones, that do not exercise the same level of control over their members. Nonetheless, Dor Yeshorim has succeeded at what it set out to do. It has created a genetic screening program that involves neither fertility limitation nor abortion.

In both forms, premarital screening and prenatal testing, prevention of Tay-Sachs disease has been successful among Ashkenazi Jews. Many explanations have been offered, over the years, about how it came to pass that the ethnic group that suffered so dreadfully from the eugenic nightmare that was Nazism nonetheless came to adopt the apparently eugenic practices of carrier identification or prenatal diagnosis or both. Some say it would not have happened had Tay-Sachs not been such a dreadful affliction, robbing little babies of their humanity and, ultimately, their lives. Others say that Jewish culture, with its traditional respect for physicians and for personal health, is responsible. Still others regard Tay-Sachs prevention as successful largely because it was private and voluntary, because there was no "big brother" either forcing people to get tested or requiring women to abort their pregnancies.[40]

Taken together, these explanations are not wrong, but they are incomplete; at least three other factors were also crucial to the success of Tay-Sachs screening. First, the screening was community based. Almost everyone involved—those who thought up screening programs, those who paid for them, those who spread the word, those who volunteered to serve, those who volunteered to be tested—was Jewish. Second, many of the people who raised the money, who did the research, who worked to set up screening programs all over the world, were people who—like those in the story of PKU—were galvanized by personal experience of the traumas of the disease and by a desire to reduce the suffering it caused. Finally, the program was pronatalist; it allowed everyone in the community to follow the biblical injunction to "be fruitful and multiply": even those who found out they were carrying a "bad" gene. To Jews of the post-Holocaust generation, the phrase "never again!" means not only that they will never forget what was done to them by the Nazis, but also that they will not lie supine if it ever begins to happen again. Getting tested for Tay-Sachs, in this context, is an act of defiance against the history of eugenic selection, not a capitulation to it.

Today, newborn screening for PKU and carrier screening for Tay-Sachs disease among Ashkenazi Jews are both well-established and uncontroversial social institutions. The first is mandated by the authority of several hundred governmental jurisdictions all over the world; the second is administered, also all over the world, by community-based organizations, by prenatal clinics and private practitioners, funded by either tax-based or corporate medical insurance. After some initial concerns, both programs have become uncontroversial—and both have grown considerably over the years. Newborn screening now tests for genetic diseases other than PKU; the so-called Jewish screening package can now locate the biochemical consequences of up to nine diseases known to have high incidence among Ashkenazi Jews.

Both programs were successful, at least in part, because they were created by members of the communities at risk: Robert Guthrie, whose son was mentally disabled and whose niece had PKU; Michael Kaback, whose own ethnic roots were Ashkenazi; Josef Ekstein and Sam and Ruth Dunkell, whose children had suffered with and died from Tay-Sachs disease. In addition, the efforts of these individuals were supported by those of community-based organizations, such as the National Association for Retarded Citizens and the Tay-Sachs Disease Association, deliberately created by relatives of the afflicted in order to lobby for programs—remediation, therapy, education, research—to assist caregivers and, if possible, to cure the afflicting conditions.

Both programs, like prenatal diagnosis, were simultaneously anti-eugenic and pronatal. By identifying newborns with PKU and alerting their parents to the need for dietary therapy, newborn screening permitted those babies to grow up more or less normally, to reach reproductive age, and to have their own children, thereby increasing the number of people who carried the gene in the next generation. By alerting every member of the family, whether afflicted or not, to the availability of prenatal diagnosis in each pregnancy, newborn screening also made it possible for the parents, the siblings, the aunts, uncles, and cousins of people with PKU to have children without fear, children who might not have been born if the screening program had not existed.

Carrier screening for Tay-Sachs has had exactly the same double effect. Before screening, heterozygous couples almost always stopped having children after they had had one child with Tay-Sachs disease, for fear of having another. After screening, coupled with prenatal diagnosis, became available, these couples had more children, and so did the siblings of afflicted children who were, themselves, carriers. The net result was an increase in the frequency of the mutated gene, albeit a very small one, in each generation.

The pronatal feature of Tay-Sachs screening, the fact that carrier couples felt encouraged to have children, was so important to the organizers of screening programs that they kept careful count of it. A chart prepared by Michael Kaback for the thirtieth anniversary of the first screening effort reveals that between 1969 and 1998, in 102 centers worldwide, 1,332,047 people were screened, 48,864 carriers were detected, 1,350 carrier couples were identified, and 3,146 pregnancies were monitored by prenatal diagnosis.[41] Kaback and his colleagues could well have stopped there, or they could have stopped with the next figure on the chart—604 fetuses who were diagnosed with Tay-Sachs—but they did not. There is one more figure, the one that matters most and that goes the furthest in explaining why Ashkenazi Jews accept carrier screening, no matter what its stigmatizing side effects may be: after monitoring with prenatal diagnosis, 2,466 "unaffected offspring" were born.

CHAPTER 5

Genetic Screening and Genocidal Claims

The story of sickle-cell disease screening in the United States is a story of failure. Unlike newborn screening for PKU and carrier screening for Tay-Sachs, carrier screening for sickle-cell disease has never been institutionalized in the United States; there are no governmental or private organizations devoted to it; the national advocacy association, the Sickle Cell Disease Association of America, does not support it (not on its website, at least); there are no special allocations of governmental funds to sustain it. To make matters worse, at least from the perspective of medical geneticists, the controversies that erupted in the 1970s over sickle-cell screening were so ugly that they continued to have a negative impact decades after the programs disappeared. Because of what was once said about sickle-cell screening, African Americans remain one of the demographic groups least likely to participate in any form of genetic testing, whether carrier testing or prenatal diagnosis, whether for sickle-cell disease or for anything else. Bitter memories, in this case, trumped concerns about individual and community health.

Sickle-cell disease was first identified by a physician in Chicago, James B. Herrick, in 1910. Like Bernard Sachs and Asbjørn Følling, Herrick was interested in applying scientific research techniques to medicine; as young man he took a year's leave from his practice to learn more about the chemistry of amino acids and proteins by working in the laboratory of the famous German chemist Emil Fischer. Shortly after returning to the United States, Herrick began to treat a young black man, a student, who was seriously anemic and complained of regular bouts of crippling pain. Herrick looked at the man's blood under a microscope, a practice that was not then

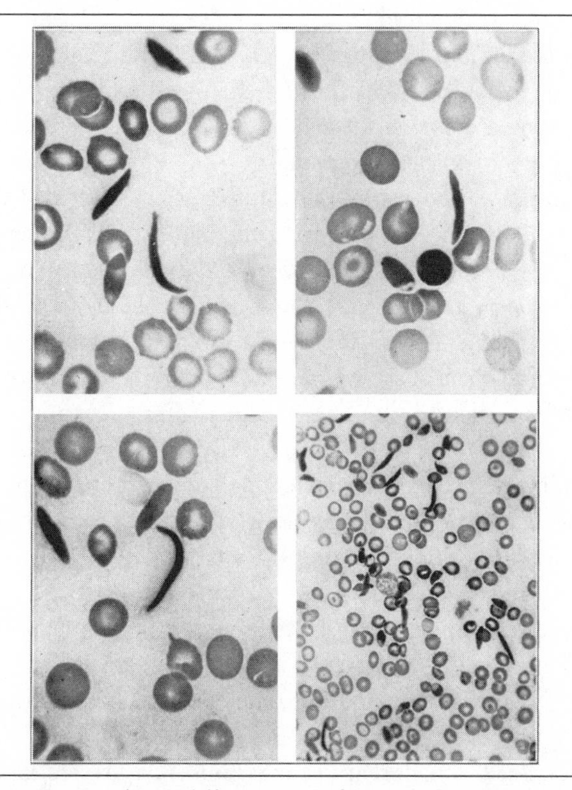

The photomicrographs, at different magnifications, that Herrick used to illustrate his case report.

From James B. Herrick, "Peculiar Elongated and Sickle-Shaped Red Blood Corpuscles in a Case of Severe Anemia," *Archives of Internal Medicine* 6 (1910): 518.

common, and discovered sickle-shaped cells scattered among the normal red blood cells.[1] In the next several years other physicians wrote similar case reports, and by the time the third and fourth reports were published the disease had acquired a name—sickle-cell anemia—and a set of diagnostic criteria: being black, becoming easily fatigued, experiencing bouts of recurrent excruciating pain, and having sickled cells in one's blood.

By the end of the 1920s two other important things had happened in the small world of research on sickle-cell disease. In 1917 an anatomist, Victor E. Emmel, developed a more reliable diagnostic test for the sickled cells. Emmel examined a black patient who had the symptoms of sickle-cell disease, but whose blood looked normal on microscopic examination. Thinking that perhaps the blood smear was in some way tainted, Emmel decided to try a slightly variant mode of preparing the blood for examination; instead of just smearing some blood on a slide, he made a ring of petroleum jelly on a sterile slide, put a drop of the blood in the ring, covered the first slide with another one (so that the blood would not dry out), and then observed the blood over the course of several hours. As time passed, some cells sickled; after a few hours, Emmel saw "a great abundance of these [sickled] structures."[2] The problem with his first smear, he concluded, had been that the blood had dried out before the cells had sickled. This new procedure (known as the "Emmel test" or "wet prep") soon became the standard diagnostic test for sickle-cell anemia.

Once he was satisfied that he had found a sound method of testing blood, Emmel also tested the blood of his patient's father, a man who had no apparent symptoms. Emmel must have suspected what many physicians subsequently came to believe: that the disease was familial, passed from parent to child. Emmel discovered that the father's blood also had a tendency to sickle if left on an airtight slide for some time. Since the father was not ill, Emmel concluded that what his diagnostic test revealed was a *latent* disease in the father, a disease that had become, for some unknown reason, *patent* in the child.

No one was able to figure out just why some people had this disease in its latent form and others had it in its patent form, but during the 1920s the conviction grew that there was a "sickling constitution" or, to use the then-current term, a "sickling diathesis," which could be inherited and which was limited, to use another contemporary term, to Negroes. Some dark-skinned people had symptoms and others did not, but if their blood cells sickled in the Emmel test, they were all thought to have the disease. Two other diseases that general practitioners dealt with in those days—tuberculosis and syphilis—behaved the same way; sometimes producing symptoms, sometimes lying in wait, even when scientifically grounded diagnostic tests revealed that the causative bacterium was present. As a consequence, it seemed rational (at least to some physician experts) to conclude that anyone who tested positive on an Emmel test was at risk for developing the often fatal and always debilitating symptoms of sickle-cell disease, even if the person, at the time of the test, had never had any bouts of the characteristic pain and was not lethargic.

By the end of the 1920s some specialists also had come to believe that anyone who tested positive on an Emmel test would inevitably pass the disease on to most of his or her offspring. After collecting pedigrees, several physicians had come to the conclusion that the disease must be a Mendelian dominant, since they had not found a single case in which a person with the patent form did not have at least one parent with the latent form.[3] Emmel testing of factory workers and schoolchildren soon led some physicians to conclude that 6 to 10 percent of American Negroes had latent cases of the disease.

Eugenic thinking was very widespread in the United States in the interwar years. Many people—including a fair number of physicians—worried that "dark-skinned" persons were destroying the "purity" of the white race by bringing "foreign diseases" across continents and oceans, slowly but surely polluting the country with "bad germs" and "bad genes." Some physicians and some eugenicists thus concluded that within a few generations all American blacks would have latent sickle-cell disease, unless something was done to keep those with "bad blood" from having children. And it would not just be blacks, unless something was done, quickly, to keep blacks from having sexual relations with whites. "Sickle cell anemia is a national health problem, especially in the United States," one southern physician warned in 1943. "Intermarriages between Negroes and white persons directly endanger the white race by transmission of the sickling trait . . . Such intermarriages, therefore, should be prohibited by federal law."[4]

All of these assumptions—that sickle-cell disease was limited to people of African descent, that the mutation that caused it was a dominant gene, and that the disease itself had latent and patent variants—were based on the Emmel test, which, by itself, could not distinguish between people with latent and patent cases of the disease. The number of physicians who believed in all of these assumptions at the time was actually quite small, largely because the number of specialists who knew or cared much about sickle-cell disease was quite small. The ordinary general practitioner was unlikely ever to see a case of the active form of the disease; many afflicted babies must have died early in their lives, and most afflicted adults were too poor to see a doctor.

Furthermore, several specialists disagreed with the eugenicists, either wholly or partially. Some argued that the latent and patent conditions were completely different diseases. Others believed the disease was not limited to Negroes, since some persons with white skin and well-established pedigrees of white ancestry had sickling red blood cells and, sometimes, symptoms of the anemia. Others recognized that the genetic reasoning was faulty, as you cannot reliably read backward from the phenotype of the offspring to the genotype of the parents. Still others argued that a gene that occurred in 6 to 10 percent of a population could not possibly be a dominant gene if it also produced, most of the time, a phenotype that was neither fatal nor particularly disabling.[5]

Unfortunately, in a culture besotted by eugenics and by a host of congruent racial stereotypes, the idea that sickle-cell disease was a dominant genotype that sometimes lay dormant (which we now know to be false) was convenient enough to gain considerable, if subterranean, currency in black and white communities alike. In black communities, the reality of the disease was terrifying: patients who had it were debilitated, partly from the persistent anemia, partly from frequent infections, and partly from regular episodes of sudden, unexplained, and excruciating pain in various parts of their bodies. These patients died young; physicians who kept track of their patients believed that few made it past twenty and none past forty. Relatives of these patients were terrified by the notion that they and their other children were susceptible; anything at all—some stress, some illness, some change in living arrangements—might bring it on at any time. Among whites, the character of the disease seemed to confirm all the worst fears about blacks: that they were lazy, that they were frequently ill, and that they could spread their illnesses widely. These stigmatizing notions would persist even after some of them were shown to be unquestionably false—and they would come to haunt the people who, with the best of intentions, tried to prevent the disease by means of genetic screening.

Uncovering the Truth

The first systematic effort to unravel the mysteries of the disease was made by James Neel; Neel's discovery of its genetic basis helped build the scientific foundation of modern medical genetics. In the early 1940s, while doing his medical residency, Neel (who also had a Ph.D. in genetics) developed an interest in thalassemia, another blood disease in which children are afflicted but their parents do not seem to be. Just before finishing his residency (and entering the armed forces), Neel coauthored a paper which argued that thalassemia *major* (the child's disease) must be a single-gene recessive (or homozygotic) condition, while thalassemia *minor* (the asymptomatic version that the parents had) must be the hybrid (or heterozygotic) condition.[6] Neel had also treated a child with sickle-cell disease during his residency. He suspected that the two diseases might be genetically similar, but he was not able to investigate the similarity until he had returned from postwar service in Japan and settled into a position at the University of Michigan in Ann Arbor.

Neel devised a double-pronged research plan. He would visit the homes of all the children in the Detroit area who had been diagnosed with sickle-cell anemia and would take blood smears not only from the children but also from their siblings, parents, and, if possible, grandparents. He would also do something novel, a comparative study; he would compare the number of sickled cells in blood smears when the children were in the crisis stage of the disease, and when they were not, with the number of sickled cells in the smears from their asymptomatic relatives. The results were straightforward: whether their disease was in crisis or not, the smears from the children contained about the same number of sickled cells—and about twice the number found in the blood of their parents and some other asymptomatic relatives. There was, as Neel put it, "a clear *hematological* distinction between the sickle cell trait [the parents] and sickle cell anemia [the children]."[7] The latent and patent forms of the disease were, Neel concluded, genetically different. As in thalassemia, he argued, the child with the patent phenotype (the symptoms, the full count of sickling cells) was a homozygote while the parents and some of the relatives (without symptoms, half counts of sickling cells) were heterozygotes or carriers.[8]

"Four months later," Neel reports in his autobiography, "there appeared an article . . . which I read with more intensity than I had ever before (or have since) accorded a scientific note."[9] This note contained the second crucial scientific discovery that unraveled the mystery of sickle-cell disease. Linus Pauling and his associates had used a then-novel analytic technique, electrophoresis, to distinguish sickling hemoglobin from normal hemoglobin. (Electrophoresis is a way of distinguishing between chemicals in a solution, based on the fact that molecules of different sizes and electric charges will migrate toward an electric pole at different rates.) They had discovered that people with sickle-cell disease had only the abnormal form of hemoglobin in their blood, but their parents and some of their other relatives had a mixture of normal and abnormal

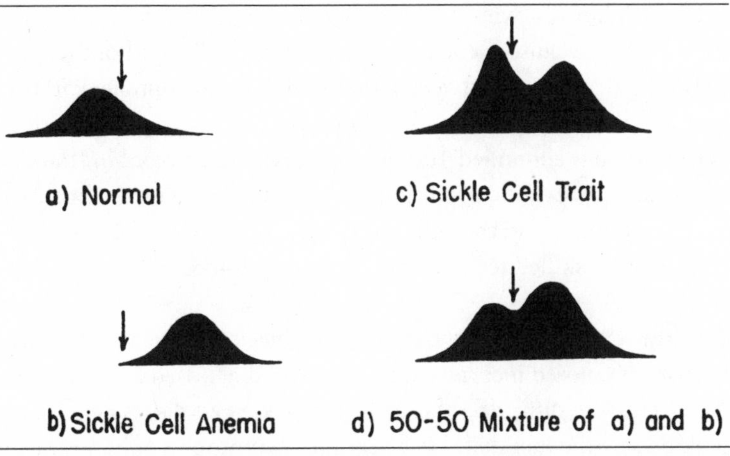

These diagrams demonstrate that normal hemoglobin (a) and sickling hemoglobin (b) have different molecular weights; the arrows are placed at a fixed molecular weight, showing that the average measurement (roughly at the top of the curve) for sickling hemoglobin is much higher than the average measurement for the normal molecule. Curves (c) and (d) demonstrate that the blood of persons with sickle-cell trait is roughly similar to a fifty-fifty mixture of normal and sickling hemoglobin.

From Linus Pauling et al., "Sickle Cell Anemia, a Molecular Disease," *Science* 110 (1949): 545. Reprinted with the permission of AAAS.

hemoglobins, half one and half the other. In other words, Pauling and his team had discovered the exact same ratio by studying types of hemoglobin molecules that Neel had found by taking a painstaking count of sickled cells in blood smears.[10]

In the space of four months, most of the old ideas about sickle-cell disease had been proven false. The disease gene was a recessive, not a dominant. Parents of children with the disease did not have latent cases of the disease. They had a completely different condition, which Neel—and, subsequently, others—called sickle-cell "trait." People with sickle-cell trait did not have to worry about getting the disease, but they did have to worry about the possibility (25 percent with each pregnancy) that their children would have the disease if their partners also had the trait.

Electrophoresis was a much better diagnostic tool than the Emmel test, because it could almost automatically, without tedious counting, distinguish between blood samples that contained 50 percent sickling hemoglobin (the trait, the heterozygotic condition) and those that contained 100 percent sickling hemoglobin (the disease, the homozygotic condition). It was also much more expensive than the Emmel test; the equipment was costly and had to be operated by very skilled technicians, but it was considerably more sensitive.

In the 1950s the chronic diseases of black Americans were frequently diagnosed incorrectly, if they were diagnosed at all. For pediatricians the difficulty of diagnosing sickle-cell disease was compounded by the fact that, in children, its symptoms could not easily be distinguished from those of several other diseases, including malaria, which was still a serious plague in some of the southern states. Many concerned physicians wanted to use the new assay to locate previously undiagnosed children, so as to be able to deliver proper treatment (antibiotics for their frequent infections; painkillers for their crises) and spare them from incorrect diagnoses and useless surgery.[11]

The creation of Medicaid in 1964 made it possible for hospitals and clinics that served economically deprived communities to improve their laboratory facilities by hiring more staff and buying equipment under special grants from state and federal agencies. Some private clinical laboratories agreed to take Medicaid payments, thus permitting concerned clinicians to attempt fairly large-scale diagnostic screening programs. As a consequence, in the late 1960s, the electrophoresis assay was being used with increasing frequency to detect new cases of sickle-cell disease. In Washington, D.C., for example, 1,251 children who were wards of the city were given complete hematological workups, including Emmel testing, in 1969. The physician in charge, Z. Ozella Webb, summed up the procedures: every child (six months to twenty-one years) was to receive "a complete blood count," and the Negro children (88 percent of the total) were also to receive a "sickle-cell prep" (the Emmel test). If the test was negative, then the children were to be retested if they showed symptoms, reached seven years of age, went into foster care, or were adopted. Positive results were to be followed up by electrophoresis. By this means the program identified one diseased child who had not been previously diagnosed and confirmed the diagnosis of four others.[12]

A Robust Screening Instrument

Despite all this progress, physicians concerned about the health of black children looked forward to a day when a simple, inexpensive technique for mass populational screening—something similar to the tuberculin test—could be used, perhaps at school entry, to locate children with undiagnosed sickle-cell disease. By the mid-1960s biochemists had figured out precisely what made sickling hemoglobin different from normal hemoglobin. Hemoglobin, they knew, is made up of two chains that have various molecules attached to them. The β chain of sickling hemoglobin, they found, has valine inserted in it where glutamic acid would normally be. More than one researcher realized that a chemical assay for valine would be both a quicker and a more reliable way of identifying sickling hemoglobin than either the Emmel test or electrophoresis. Sometime in 1967 or early 1968 the Ortho Pharmaceutical Company sent samples of one such test out to researchers for clinical trials; they called it Sickledex.

Sickledex kits consisted of a powder, a liquid, a test tube, and a piece of printed cardboard. To use the kit, one dissolved the powder into the solution in the test tube and then added a drop of blood (blood from a pricked finger would do). After two to five minutes, one held the test tube up against the card. If the solution was cloudy, sickling hemoglobin was present; if it was transparent and pink, sickling hemoglobin was absent. No special skills were needed; neither a microscope nor an electrophoresis assembly was required. Like the Guthrie card and the automated Hex-A assay, Sickledex could be used for mass screening of at-risk individuals: if it was accurate, anyone would be able to use it, almost anywhere, at almost any time. Early tests in the laboratories of sickle-cell researchers indicated a low false-positive rate. "Sickledex is quicker to perform than the orthodox sickling test [the Emmel test]," one laboratory reported, "and is more reliable in inexperienced hands."[13]

Sickledex did have one failing, however. Like the Emmel test, it revealed the presence of sickling hemoglobin, but it did not reveal whether the blood being tested came from someone who was a carrier (50 percent sickling hemoglobin) or from someone who had the disease (100 percent sickling hemoglobin); only electrophoresis could do that. As a consequence, populational screening, if it was instituted, would have to be a two-step process: first the robust screen, Sickledex, to find out whether sickling hemoglobin was present, then, for those who tested positive with Sickledex, the sensitive screen, electrophoresis, to find out how much. Proponents of populational screening for sickle-cell disease initially thought this was a very good thing; the double test would serve a double goal, identifying people with previously undiagnosed cases of the disease at the same time as it identified people who were carriers, people who might want to exercise reproductive caution. These optimists turned out to be wrong; the double purpose of the double test was precisely the technological barrier at which mass screening for sickle-cell anemia stumbled.

Organizing the Screening Programs

For a variety of reasons, by the early 1970s there were a fair number of proponents of screening, influential people who were ready, willing, and indeed anxious to take up the challenge of mass screening, both for sickle-cell disease and for carrier status. The science seemed to be in place, the technology was reasonably priced, and expectations were high. A cure for sickle-cell disease might be found, but even before that the disease might be remediable, like PKU, or it might be preventable, like Tay-Sachs. In several cities black professionals were organizing philanthropic organizations, modeled after the National Association for the Advancement of Colored People (the NAACP) and the March of Dimes, intending to raise money for and consciousness about the disease. One of the first of these was the (New York) Foundation for Research and Education in Sickle Cell Disease, which was "composed of individuals whose professional activities in medicine, education, law, and religion are directly concerned with this severe problem of the black community." The foundation had four goals: to encourage research leading to better therapies and a cure; to communicate information about the disease to medical professionals so as to ensure that "diagnostic tests for sickle cell disease [are] a regular part of diagnostic procedures"; to communicate information about the disease to the general public "to aid in case finding and acceptance and understanding of the affected person and family"; and to encourage "premarital genetic counseling" so that "every black person should know whether they carry the sickling condition or not, and if so, in what form."[14]

Similar organizations had sprung up, at about the same time, in cities as diverse as Chicago, Los Angeles, Memphis, Richmond, and Philadelphia. All were anxious to raise money and awareness and most seem to have favored, as the prospectus for the Foundation for Research and Education put it, "universal testing of all black individuals, as well as all individuals from other areas of high incidence."[15] Initially, progress was slow but steady. A few school districts began testing their children; brochures were printed and delivered to public health centers, churches, and professional meetings; and several hospitals created special clinics.

In October 1970 the efforts of these philanthropic organizations received a significant boost when a hematologist on the faculty of the Medical College of Virginia, Robert B. Scott, published a "special communication" in the *Journal of the American Medical Association (JAMA)*. Scott made what might be called a civil rights argument for increased funding for sickle-cell disease. First he demonstrated that the incidence of the disease was much higher than that of such diseases as cystic fibrosis, muscular dystrophy, and PKU. Then he demonstrated (through analysis of the monies raised by volunteer associations and the number of grants funded by NIH) that support for research on sickle-cell disease, despite its high incidence, lagged significantly behind support for all the other diseases. For reasons of both justice and public health, he argued for a change of national priorities. His argument was seconded by the editors of *JAMA,* who opined, in an editorial in the same issue, that the current situation was a "public failure," and that righting it would require "expanded programs of comprehensive care, including diagnosis, treatment, and rehabilitation with emphasis on mass screening, genetic counseling, and public education." Scott and the editors were arguing, in short, for the double test with a double goal: identifying carriers and identifying the afflicted.[16]

The very next day the *New York Times* covered the publication of the article in a story entitled "Doctor asks Curb of Negro Disease." The *New England Journal of Medicine* republished the piece a few weeks later. *Time, Newsweek,* and *Ebony* also published articles about sickle-cell disease within weeks. Scott referred to Sickledex in his original article as "a simple test for sickle trait carriers which makes screening possible"; he hoped—as did the editors of *JAMA* and the leaders of black philanthropic associations—that Sickledex would make it possible to identify new cases of the disease among the nation's children, but also, and at the same time, turn sickle-cell disease into "the first hereditary illness which could be controlled by genetic counseling."[17]

In short order, sickle-cell anemia became what might well be called the celebrity disease of the early 1970s. Black civil rights activists, having scored significant successes both in the domain of political liberties (voting rights) and education (school desegregation), had been turning to economic issues since the mid-1960s. The poor health of black Americans and the poor quality of health care in black communities were among those issues. Sickle-cell disease, although far from the most significant factor in black mortality and morbidity, was one disease about which a great deal was known and, so it seemed, about which a great deal could be done, quite expediently. Black celebrities were ready and willing to lend a hand: the boxers Muhammad Ali and Joe Frazier advocated better public information about the disease; so did the baseball player "Doc" Ellis (who was a carrier). Actors like Bill Cosby and Sidney Poitier starred in movies that featured the disease in their plot lines. The musicians Louis Armstrong and Dizzy Gillespie played numerous benefit performances; their wives became chairs and co-chairs of fundraising committees. An episode of the popular television program *Marcus Welby, M.D.* was devoted to the sufferings of a sickle-cell patient.

In the fall and winter of 1970 the political ball had also begun to roll. Staff members at the U.S. Department of Health, Education, and Welfare (HEW) were preparing materials for President Nixon's upcoming message to Congress about health. A young HEW fellow, Colby King, had written a report on sickle-cell disease, which he passed on to his supervisor, Robert Petricelli, who recommended that increased funding for the disease be part of the president's health message. Petricelli also gave the report to his father, Leonard Petricelli, who owned a television station in Hartford, Connecticut. By early November 1970, just weeks after Robert Scott's article had appeared in *JAMA,* the elder Petricelli had aired the first of four local television specials focused on sickle-cell disease and had started collecting funds to sponsor research on the disease being done at Howard University. Within months, largely through Petricelli's efforts, Connecticut had passed legislation to fund sickle-cell screening for all children in the state in grades seven through twelve. Subsequently the Hartford school district sent letters to the parents of black children, asking for permission to perform the test. The program was endorsed by several local physicians and black community groups, and a large proportion of the parents agreed to have their children tested.[18]

Within eighteen months, twelve states and the District of Columbia had followed Connecticut's example, passing laws related to sickle-cell screening. President Nixon had, indeed, highlighted sickle-cell disease in his spring message to Congress. Later that year both houses passed an amendment to the Social Security Act (called Early and Periodic Screening, Diagnosis and Treatment Programs) requiring every state to offer screening for sickle-cell disease to children on Medicaid. A year later Congress passed the National Sickle-cell Anemia Control Act, which required HEW to provide funding for screening, to be administered through hospital-based comprehensive-care programs.

Many of these enactments had been proposed by black legislators and had been passed—this being the period immediately after the successes of the civil rights movement—with very little objection, almost no discussion, and the best of intentions.[19] Everyone involved seemed to think, as President Nixon had put it, that something had to be done quickly to make up for the decades during which sickle-cell disease had been an invisible malady.

Unfortunately, the provisions of the state laws varied widely, and many confused the goal of identifying afflicted children with the goal of identifying carriers. Connecticut permitted and funded screening of children through school districts but did not mandate it; Massachusetts and nine other states mandated it for all children entering public schools but did not fund it. Arizona allowed Medicaid funding for carrier testing during prenatal care, without addressing the need for disease screening. Georgia and Kentucky required and funded testing of newborns in hospitals; Louisiana encouraged it but did not provide funding. (Blood testing of newborns turned out to be unreliable because their blood-forming tissues are still producing fetal hemoglobin.) Nine states insisted that couples could not receive marriage licenses unless they had been tested; one state, Virginia, mandated testing of all persons in state institutions and of all children at the age of six. Laws in only two states directly targeted black people; Kentucky, for example, specified that newborn Negro children and Negro couples applying for marriage licenses had to be tested. In other states the targeting was done indirectly. New York required testing of all children in "city school districts" and all couples who presented themselves for marriage licenses and were not of the "Caucasian, Indian or Oriental races." Only three states specifically authorized special educational programs about sickle-cell disease, and they were not the same three states that authorized genetic counseling as part of the testing program. The test results were to be kept confidential in only one state, Maryland.[20]

Unanticipated problems arose almost immediately. In some communities black civic leaders objected to the testing programs as stigmatizing and refused to lend their support. In other communities black leaders objected to the testing programs because they were mandated. Howard Pearson, a physician at Yale University Medical School, told a reporter that in New Haven, "people apparently felt that screening was being forced on them, and they wouldn't go along with it."[21] State coercion did not sit lightly on the shoulders of black communities, just recently liberated from being forced to use separate bathrooms, drinking fountains, and schools. In states that mandated premarital screening, couples applying for marriage licenses suddenly found that in order to get married they had to go to a medical laboratory and pay for a blood test. Some civil rights advocates, black and white, argued that their state's law was discriminatory and therefore unconstitutional. Sometimes community-based clinics and private diagnostic laboratories mistakenly tried to save money by using Sickledex alone or even the old Emmel test, which meant that numerous people (children, teenagers, and even adults) were told that they had the disease when they were in fact carriers. When small children were properly tested in comprehensive-care centers, the results were still a disappointment to the organizers of the programs, because follow-up with the mothers proved extremely difficult. One well-managed program in a teaching hospital in upstate New York was unable locate 25 percent of the mothers for follow-up testing of their babies just one month after their birth.[22]

Given time, all these problems could probably have been overcome, just as they already had been for PKU testing and were, in the very same years, for Tay-Sachs screening. But two profoundly serious sets of complexities—one set medical, the other political—meant that sickle-cell screening became very controversial very quickly, leaving no time for writing new laws, educating members of the public, or restructuring testing regimens.

Medical Controversies

The medical complexity had to do with the confusion about the difference between having sickle-cell trait and having sickle-cell disease; between being a heterozygote and being a homozygote. In PKU testing this distinction was initially beside the point. The Guthrie test could only reveal afflicted newborns, but that was all that was wanted at the time. In Tay-Sachs screening the distinction was also beside the point, but the point was different. Homozygosity could be diagnosed by observing the afflicted child; mass screening was meant to reveal carrier status, heterozygosity. But testing for sickle-cell disease or sickle-cell trait was, of necessity, a two-stage process. Both stages had to be completed to reveal who was afflicted with the disease, but, inevitably, carrier status would also be reported: if you passed the first stage, went on to the second, but turned out not to be a homozygote, you were a carrier. Unlike the Guthrie test and the Hex-A assay, in the case of sickle-cell disease the search for homozygotes was bound to reveal who was a carrier and who was not.

In the early 1970s many educated people misunderstood Mendelian genetics in general and its application to single-gene recessive conditions in particular. Just one example: the very first line of the National Sickle-cell Anemia Control Act says that "sickle cell anemia is a debilitating, inheritable disease that affects approximately two million American citizens," a statement which confuses the number of persons who were then assumed to be carriers with the number who were then assumed to have the disease. To make matters worse, many physicians and nurses who got their training before 1970, if they learned anything at all about sickle-cell disease, could well have been taught the old doctrine that everyone who had the trait—that is, everyone who tested positive on an Emmel test—was at risk, since latent cases might turn into patent cases overnight. A 1974 study found that one in five physicians could not distinguish sickle-cell trait from sickle-cell disease.[23]

In addition, in the early 1970s there was an ongoing and very acrimonious debate among physician specialists (who, by then, perfectly well understood the underlying genetic phenomena) about the health status of sickle-cell carriers. Some physicians had feared for years that if half of a carrier's hemoglobin would sickle, the trait could not possibly be altogether benign: "Occasional reports," several physicians wrote in 1970, ". . . indicate that in certain unfavorable circumstances these patients may experience the severe vaso-occlusive episodes characteristic of homozygous sickle-cell anemia." One of those reports had appeared in the fall of 1968, after the Summer Olympic Games held in Mexico City: "Athletes with sickle cell trait were having difficulties because of the lower oxygen tension of high altitudes."[24] In the spring of 1970, furthermore, an article appeared in the prestigious *New England Journal of Medicine* arguing that the deaths of four young black men during basic training at a high-altitude army post were the result of sickle-cell trait. A fierce controversy ensued. Medical opinion was divided about whether these disabilities and deaths were correctly attributed to sickle-cell trait; there were, some specialists argued, too few cases to move from correlation to cause. The medical argument did not, however, prevent a black serviceman, James A. Powell, from suing the secretary of the army, asking that he and all other carriers receive honorable discharges and that retention of soldiers with the trait be declared illegal.[25]

The U.S. Department of Defense, erring on the side of caution, issued orders restricting the activities of soldiers who were known carriers of the trait, including, but not limited to, prohibiting them from serving as pilots or deep-sea divers.[26] The controversy led the

Department of Defense to pay the National Academy of Sciences (NAS) to undertake a study of the issue and make recommendations. In 1973 the NAS panel, chaired by Dr. Robert Murray, a hematologist at Howard University Medical School, recommended that *all* recruits for *all* the armed forces be screened for sickle-cell disease and hemoglobin disorders, but that the restrictions on activities be lifted, except the prohibitions on serving as pilots and deep sea divers. The Air Force Academy, following the recommendations of this panel, decided to exclude applicants with sickle-cell trait and to force cadets who were carriers to resign (with an honorable discharge)—a set of policies that was not reversed until several years later, when one of the cadets, assisted by dissenting physicians and several civil rights organizations, sued. Reflecting the lack of medical consensus on this issue, the suit (which was settled out of court) argued that there was no scientific basis for these discriminatory policies, despite the fact that the original recommendations had been made by a committee of the NAS.[27]

When medical experts cannot agree and when many physicians hold outdated ideas, it is not surprising that, when faced with decisions about health care, members of the nonmedical public have difficulty sorting out "right" from "wrong." In the early 1970s there was a great deal of confusion and misinformation in the communities targeted by testing. Many mothers, for example, when notified that their children were "carriers of sickle-cell trait," refused to believe that their children were not sick. One study of sixty-seven Seattle mothers whose children had tested positive for the trait revealed that 43 percent of them thought that their children had a disease and 72 percent of them had put some restrictions on their children's activities, in some cases completely prohibiting athletic activity. More than half of the mothers in the Seattle study also thought that their children would benefit from supplementary iron in their diets. "Sickle-cell *non*disease," the researchers concluded, was an epidemic *produced* by screening programs, and, they argued, it was becoming even more stigmatizing than the disease itself.[28]

Some insurance companies also acted as if sickle-cell trait were a disease in and of itself and raised their rates (on both life and health insurance) for people whose tests revealed that they were carriers. In 1972 Joseph Christian, a sickle-cell researcher at the Indiana University School of Medicine, asserted that several insurance companies "charge as much as 150 percent of the usual premium, some added 30 percent" to insure those with the sickle-cell trait.[29] In addition, reports abounded of stewardesses and pilots who were fired when their employers discovered that they were carriers and of companies that required testing as a condition of being hired. Although little evidence seems to have been found to substantiate those reports, the net effect of the rumors was substantial; in one workplace after another (including the NIH itself and several Veterans Administration hospitals) workers refused to be tested, fearing that their jobs and their health insurance would be jeopardized.[30]

The health consequences of carrier status were not the only subject on which sickle-cell experts disagreed. They also argued about health priorities and, as a consequence, about what was the "right" moment in the life cycle at which to provide screening. Like several of the other disputes, this argument arose from the intersection of the technologies of testing and the social characteristics of the population being tested. If the primary goal of a testing program was to identify previously undiagnosed children, then public school entrance seemed an effective intervention point, just as it then was for tuberculin testing and polio immunization. All experts agreed that if the child was a homozygote, the parents should be notified.

But on that point agreement ended. Should the parents, the obligate heterozygotes, also be given reproductive counseling? Perhaps not, some argued: the counseling might so distress them that they would refuse further contact with the system that had brought the unwelcome news. Absolutely yes, others maintained. Especially after the details of the Tuskegee syphilis study were revealed in early 1972, many people came to believe that health information should not be kept hidden from anyone, no matter what the consequences. And what if the child turned out to be a heterozygote, a carrier? Do you give that information to a five-year-old? Obviously not. But should you give it to a hard-pressed parent, who is unlikely either to remember it or to control the child's sexual activity a decade later? Indeed, if that hard-pressed parent is a poverty-stricken single mother, is she likely to act on the information and get herself—let alone her sexual partner—tested?

By the mid-1970s, as a result of all these complexities and disagreements, the early enthusiasm for screening for sickle-cell disease had been dampened considerably. The pioneers had come to realize that their desire to use mass screening to diagnose afflicted children was doomed to failure, partly because follow-up diagnosis and treatment was endangered by the screening process itself. Withholding the inevitably available information about carrier status was either futile or unthinkable in the new age of informed consent, but providing the information could be unnecessarily stigmatizing, utterly useless, or, worse, medically counterproductive.

Political Controversies

If the first goal of mass screening, identifying previously undiagnosed cases of sickle-cell disease, was controversial, the second goal, prevention of the disease through identification of carriers and reproductive counseling, was more so. "Reproductive counseling" was a euphemism for reproductive restraint, birth control, and abortion—and these were all hot-button divisive issues in the black communities of the early 1970s.

African-American objections to reproductive counseling were

multifaceted and mutually reinforcing; many harked back to the early history of eugenics in the United States. The black debate about birth control had started early in the twentieth century, just as the birth control movement itself was gaining momentum. In the 1920s W. E. B. Du Bois, one of the founders of the NAACP, had published ringing endorsements of Margaret Sanger's efforts to make birth control devices and information widely available. Du Bois believed that unfettered reproduction led to horribly high infant mortality rates, destroyed women's health, and kept black families in poverty. But many of Sanger's supporters had made public their support for eugenics, arguing that birth control was the best way to keep the "inferior" races from outbreeding their "betters," and many black leaders came to believe that, on this subject, Du Bois had been duped. In the 1930s Marcus Garvey's Universal Negro Improvement Association, an early black nationalist movement, had condemned birth control, not only as being contrary to nature but also as a white plot to drive the black race into extinction.[31]

The eugenic implications of birth control and the "quality-versus-quantity" debate that ensued had made it exceedingly difficult to open birth control clinics in black communities during the first half of the twentieth century, and they remained a potent impediment decades later.[32] In 1967 an amendment to the Social Security Act required the federal government to spend 6 percent of the funds for Maternal and Child Health Services on family planning programs. In the aftermath of the civil rights movement, however, some young black activists had become radicalized—and their particular brand of radicalism led them to agree with Marcus Garvey on the subject of family planning. "Our safety, our survival, literally depend on our ever increasing numbers," one wrote in the *Liberator* in 1969. As a consequence, in the late 1960s and early 1970s, when some black community leaders succeeded in opening family planning clinics in black neighborhoods, others organized picket lines in an effort to shut the clinics down. Such arguments had a powerful appeal. "Blacks must be suspicious [of all population-control programs]," the Congressional Black Caucus asserted in the spring of 1972, "since Blacks are the victims of racists and white supremacists who will do anything in their power to maintain white supremacy . . . [and] family planning is no exception to this." When the comedian Dick Gregory published an article in the mainstream black magazine *Ebony,* prescribing big families as a protest against white plans for genocide, some readers wrote in to congratulate him for "telling it like it is."[33]

Many black women, and probably most black feminists, disagreed with this pervasive antipathy to birth control. A fair number of the *Ebony* letter writers who disagreed with Gregory were women. "If all of us blacks had the money [Dick Gregory] has, we would have no excuse for not having larger families," one wrote. "[But] the majority of us live on smaller incomes . . . The female who bears children must think of herself—physically, financially and mentally." Another writer was more pointed: "[Gregory] is using the body of his wife as a weapon, reducing her to the level of a cow, and she apparently accepts her lot as inevitable . . . I would implore Mr. Gregory to fight his war and his bitterness as a man, and thus allow his wife to become a person." Shirley Chisholm, the first black woman elected to Congress, put the matter succinctly in her 1970 autobiography: "To label family planning and legal abortion programs 'genocide' is male rhetoric, for male ears. It falls flat to female listeners . . . Women know . . . that two or three children who are wanted . . . will mean more for the future of the black and brown races . . . than any number of neglected, hungry, ill-housed and ill-clothed youngsters. Poor women of every race feel as I do."[34]

In part because of the political turmoil stimulated by publication of the controversial Moynihan Report in 1965, what Chisholm regarded as "male rhetoric" tended to dominate black discussions in those years. The report, actually entitled *The Negro Family:*

The Case for National Action, was prepared by Daniel Patrick Moynihan (later to become a senator from New York), who was then an assistant secretary in the Department of Labor. Moynihan had found data documenting what he interpreted as the "collapse of the black family." Three pieces of data were crucial to his argument: (1) in urban areas, husbands were either absent or divorced in 22.9 percent of nonwhite families but in only 7.9 percent of white families; (2) in the northeastern states, 26 percent of once-married black women either were divorced or had absent husbands; and (3) between 1940 and 1960 out-of-wedlock births had increased from 16.8 to 23.6 percent in the black population but only from 2 to 3.07 percent in the white population.

Many liberals, both black and white, objected to the Moynihan Report—not to the data but to his interpretation of them. Moynihan argued that slavery had cut the cultural ties that bound families together and that the current regime of welfare was making matters worse by refusing assistance to intact families in which the father was either employed or capable of employment. Moynihan saw this welfare regime as encouraging black women not to marry and not to reside with the fathers of their children. Liberals saw things differently: they saw the root cause of family disintegration as poverty, not welfare payments; they saw the black family as a different kind of cultural unit from the white family, but equally deserving of respect; they understood Moynihan to be "blaming the victims"; in short, they viewed him as unable to understand or sympathize with black cultural norms, and they interpreted this inability as racism, pure and simple.

Publication of the Moynihan Report thus strengthened black opposition to birth control, abortion, voluntary sterilization, and abstinence: the entire array of family planning techniques. After 1965 black leaders who argued in favor of family planning ran the risk of being called racists by their peers. And, of course, no matter what euphemisms they used, proponents of screening for carriers of sickle-cell disease ran the same risk; there was no hiding the fact that they could not achieve "prevention" or "control" without urging carriers to "reproduce responsibly."

In the fall of 1973, for example, Alyce C. Gullattee, a psychiatrist on the faculty of Howard University College of Medicine, writing in the *Journal of the National Medical Association,* a publication aimed at black physicians, struck a note that was repeatedly heard in the next few years. We have all been deceived by the "altruism" of governmental bodies, Gullattee declared sardonically:

> We are so deeply concerned with sickle cell anemia because of: (1) supply and demand—the need for a captive labor force that is not competitive; (2) population preoccupation and selective euthanasia; (3) *eugenics and legislatively sanctioned genocide;* and (4) intellectual diversion of the less perceptive through gifts, such as government funds for sickle cell anemia that are cloaked with subtle white supremacy.[35]

The Moynihan Report had thus not only helped resurrect the old, now outdated connection between family planning and eugenics, it had also caused many black activists to perceive the proponents of sickle-cell screening as both racist and genocidal.

For one group of black radicals in the 1970s, creating the perception that sickle-cell screening was racist and genocidal was a deliberate political strategy. Like many advocates of "black power" of that era, the Black Panthers considered moderate, middle-class African Americans—the kind of people who were proponents of sickle-cell screening—insufficiently revolutionary. The Panthers scorned the nonviolent ethos of Martin Luther King Jr.; they ridiculed the NAACP and the United Negro College Fund ("Negro" was a word they refused to use); and they dismissed school desegregation and voting rights as sops to white consciences that did nothing to alleviate the economic disadvantages with which black people struggled. They changed their names, their hairstyles, and their clothing—all to reflect their desire to separate from, not assimilate into, white American culture. They referred to black moderates derisively as "Oreos" (black on the outside, but white on the inside); one of the "Oreo" cultural practices that they despised the most was "reproductive responsibility," the maintenance of small and stable families.

The Panthers believed that moderate blacks, people who had "made it," were ignoring the interests of the majority of blacks ("The People") who were poor and, in their poverty, disenfranchised. They heaped particular scorn on those black professionals who tried to alleviate the problems of the poor through governmentally funded institutions: schools, public health departments, hospital clinics, and welfare programs. They wanted to return "power to the people" by creating alternative institutions in which the needs of the poor could be met by the poor themselves, working in tandem with sympathetic radical professionals, people referred to as "indigenous consumers and providers."[36]

So the Panthers set about creating "survival programs" in their own community centers in major American cities: free breakfasts for schoolchildren, free bus services for families to visit imprisoned relatives, free courses in various subjects, and free medical checkups conducted by volunteer physicians and nurses. These programs were intended to "insure the continued existence of the people . . . by meeting their basic needs, such as food, clothing, medical care and revolutionary education." They had to be funded and operated by the party and the people in order to "combat and overcome the plots of genocide implemented by the oppressor."[37] By the spring of 1971 survival programs of one sort or another had been implemented in Panther centers in Oakland, Chicago, New Haven, Nashville, Boston, Philadelphia, and Los Angeles.

Then, that spring, just after President Nixon's speech to Congress brought sickle-cell disease into the national spotlight, the Panthers added free Sickledex testing to their roster of services. "You can

HELP DESTROY ONE OF THE ATTEMPTS TO COMMIT BLACK GENOCIDE," their newspaper proclaimed, "FIGHT SICKLE CELL ANEMIA!" "Sickle cell anemia," the article continued, "is a deadly blood disease that is peculiar to black people . . . The racist U.S. power structure has no intention of ceasing this form of genocide, since it is this racist power structure that perpetuates this disease."[38]

The Panthers wanted The People to fight sickle-cell anemia by

Members of the Black Panther Party picketing in Oakland, California, in 1971.

From Stephen Shames, *The Black Panthers: Photographs* (2006), with the permission of Stephen Shames.

getting Sickledex tests in Panther clinics and donating money to The People's Sickle Cell Anemia Fund. All other testing programs, they argued, were funded by governments—local, state, federal—and therefore suspect. All other foundations were "PHONEY FOUNDATIONS TRY[ING] TO SABOTAGE BLACK PANTHER PARTY'S SICKLE CELL PROGRAM," as one headline declared. The text of the accompanying article complained that these organizations—the Los Angeles Sickle Cell Anemia Foundation, the National Sickle Cell Disease Research Foundation in New York, and the National Institutes of Health—were "duping the people, nationwide, out of hundreds of thousands of dollars," to establish screening programs that were ineffective, "offering hope when they know they never intend to deliver."[39] The board members of these foundations, prominent black professionals and celebrities, were, according to the Panthers, "scum."

At every possible opportunity, the Panthers tried to reinforce the notion that sickle-cell programs funded by governments had to be genocidal. Photographs of red blood cells were accompanied by a headline that read "GENOCIDE: THE SYSTEMATIC KILLING OR EXTERMINATION OF A WHOLE PEOPLE," while the text asserted that the U.S. government knew that sickle-cell disease was fatal and was deliberately refusing to do research on a cure. Furthermore, the Panthers argued, the screening programs in comprehensive-care centers funded by the government were nothing more than efforts to distract attention from the ultimate, genocidal goal. Another article, condemning the reproductive counseling offered by non-Panther testing programs, declared: "SICKLE CELL ANEMIA + the 'pill' = instant genocide."[40]

The reproductive goals of sickle-cell screening programs were thus one of the many battlefields on which black Americans of the early 1970s struggled with one another. Nothing illustrates this corrosive effect better than an incident that occurred in April 1974. The Los Angeles Sickle Cell Anemia Foundation held a fundraising dinner dance in a large hotel in Beverly Hills. The Los Angeles branch of the Black Panther Party set up picket lines in front of the hotel, attempting to prevent the guests from entering. An officer of the foundation called the Los Angeles police and demanded that the pickets be arrested. Several members of the Party spent the night in jail.[41]

By 1980 virtually all state mandates for mass screening for sickle-cell disease had been rescinded and the screening programs organized by government-funded comprehensive-care centers were closed. The failure of large-scale community-based sickle-cell screening in the early 1970s had probably been, as social scientists like to say, overdetermined. Nothing went right; every unexpected outcome was negative. The testing process was very complicated and could not easily be explained to the at-risk population. The various mandates for testing were well intentioned but ill conceived, and in their ill conception irritated the very raw nerves of leaders of the population the mandates were meant to serve. The experts did not speak with one voice. They disagreed about the health implications of carrier status; they disagreed about how best to counsel carriers; and they disagreed about how best to construct a mass-testing enterprise. The socioeconomic characteristics of the at-risk population made follow-up extremely difficult, whether for therapeutic intervention or for reproductive counseling—and when counseling could be achieved, neither comprehension nor cooperation could be ensured.

Most detrimental of all, the political and intellectual leaders of black communities came, very quickly, to oppose screening, even those who had been initially supportive. Some feared stigmatization, a fear not to be taken lightly. Others feared governmental mandates, a potent fear in a population that had suffered the mandates of racial segregation for almost a century.

The most damaging fear of all was the fear that the effort to prevent sickle-cell disease by a combination of diagnostic technology and reproductive restraint would result in genocide. Had the governments, state and federal, not become involved, this fear might have been overcome. But the governments did become involved, from the very beginning, and black distrust of government prevailed. Had black philanthropic organizations of the early 1970s been united behind mass screening, that distrust might have been overcome. But far from being united, the kinds of philanthropic organizations that had been critical to the triumph of screening for Tay-Sachs and PKU were, in the case of sickle-cell disease, locked in battle with one another.

Today, testing for the sickle-cell trait is conducted, pregnant patient by pregnant patient, as part of prenatal care, but (in the United States, at least) subsequent partner screening, prenatal diagnosis, and abortion are not nearly as widespread as they are for Tay-Sachs disease in the Jewish community. Fears of eugenics die slowly; many African Americans, young and old, male and female, reject the services of medical geneticists (even when the condition at issue is not sickle-cell disease) on the grounds that medical genetics is eugenic, yet another instance of white efforts to control and limit their fertility.[42] With regard to screening for sickle-cell disease, the memory of what had once been done (and said) in the name of eugenics proved to be more important than the reality of painful, debilitating disease. In the end, the eugenic *non*implications of carrier screening turned out to be as potent and as stigmatizing as those of sickle-cell *non*disease.

CHAPTER 6

Parents, Politicians, Physicians, and Priests

However effective genetic screening may be at preventing serious diseases, very few recent governments have been willing to mandate it. Governments sometimes require quarantines, to prevent the spread of epidemic disease, and they often require immunization, to prevent the appearance of diseases which might become epidemic—but they usually do not require genetic screening, even though it might be just as effective a way to reduce the toll that illness takes on a population. No government official wants to go down in history as advocating something the Nazis once advocated; no government, especially a government elected after World War II, wants to be tarred with the brush of eugenics.

Except for two—both of them on the Mediterranean island of Cyprus. In the Republic of Cyprus (otherwise known as Greek Cyprus) and also in the Turkish Republic of Northern Cyprus, every person who wants to get married must submit to a blood test to determine whether she or he is a carrier of thalassemia—or to put it more accurately, a carrier of one of the several mutations of the β-globin gene—before a license will be issued. Strictly speaking, only one of these programs is entirely a governmental mandate. The one in the Turkish Republic of Northern Cyprus was created by parliamentary legislation and receives its funding from the ministry of health. The screening program in Greek Cyprus, however, although operated by the government, was actually created by the Cypriot Orthodox Church. Cypriot Orthodox priests ask brides and grooms to present certificates attesting to screening before marriage ceremonies are scheduled; virtually all couples comply. Since the majority of Greek Cypriot marriages are consecrated in the church, thalassemia screening reaches a very large proportion of the population.

Both Cypriot republics are democratically governed. Since both genetic screening programs have been operating since the early 1980s, at least two generations of young Cypriots have had the blood test thrust upon them. No Cypriot can be unaware that screening is going on; both programs operate out of buildings that are prominently situated and labeled. Yet neither country has experienced any appreciable public protest against mandated screening. Cypriots appear to accept genetic screening for thalassemia as a reasonable public health measure, sanctioned not only by medical experts but also, in the Greek part of the island, by the established church, its most respected cultural and educational institution. Some outsiders have accused the Cypriots of being pretty far down on the slippery slope of eugenics, but the Cypriots themselves do not agree—for reasons that are worth exploring in some detail.

Cyprus lies in the eastern Mediterranean, south of Turkey and north of Egypt, at a geographical and historical junction between Asia and Europe. It is the third-largest island in the Mediterranean, 9,250 square kilometers, a little more than half the size of the state of Connecticut. The island was once heavily forested and rich in copper, which meant that it was considered very valuable real estate, not infrequently passed from one dominating empire to another: Phoenician, Hittite, Hellenistic, Persian, Ptolemaic, Roman, Byzantine, Venetian, Ottoman. The population (roughly one million in 2006) has been split, for many centuries, between people of

Greek heritage (currently about 77 percent of the population) and those of Turkish heritage (about 18 percent).

In the 1880s the British took over administration of the island from the Turks in exchange for a portion of the tax revenues; the Ottoman Empire was on a downward slide into bankruptcy, and the British needed a naval outpost with which to protect the newly opened Suez Canal. When British sovereignty began, the vast majority of Cypriots were agricultural workers, either on their own land or on someone else's. A very few families had become wealthy and educated, either as (mostly Turkish) governmental functionaries, in trade (both Greeks and Turks), or as members of the Orthodox Church hierarchy (unlike Roman Catholics, Orthodox clergy may marry). Tradition has it that, wealthy or poor, Turkish and Greek Cypriots lived together as neighbors, albeit not exactly friendly neighbors: they neither spoke the same language nor practiced the same religion, and they did not intermarry. Traditions are, however, sometimes semimythical; the Greek-speaking and Turkish-speaking residents of Cyprus probably did not intermarry, but they certainly had sexual relations. Recent DNA studies reveal that in their genetic makeup the Greek and Turkish Cypriots are more similar to each other than to either of the populations from which they originated.[1]

As it did in its other colonial outposts, Britain soon started building a civil infrastructure on Cyprus. Roads were laid and maintained, English schools were constructed and staffed, and drainage ditches and water purification facilities were built. Medical clinics were also opened, and physicians and nurses were sent from Britain to staff them; eventually there would be a modern multiwinged hospital for the capital, and several smaller hospitals and clinics in the port towns. Cyprus became a British crown colony in 1925 as part of the spoils of World War I, and at that point British civilians began to arrive in some numbers, not only as tourists, but also as the owners and managers of businesses of all sorts. Cyprus was closer to Britain than its other major colonies in India or Australia, opportunities abounded there—and the weather was delightful.

The British, like all colonial powers, built this infrastructure in their own interests, but a not unexpected outcome was that living conditions began to change for the Cypriots as well. Purified water meant better health for everyone who drank it, while roads made transportation easier for everyone who used them. Some Cypriot children, both Turkish and Greek, were sent to British schools and, having learned English, found employment either with the imperial government or in British businesses. Some Cypriot businesses expanded, and some wealthy landowners became even wealthier supplying the needs of the British and their naval bases. The Cypriot population of the capital and the port towns grew as Cypriots increasingly began to leave their villages to find work in the cities.

During World War II, Cyprus flourished, as the naval bases were in constant use and the island was neither captured nor attacked. After 1945 a fair number of educated Cypriot youth traveled to the "home country" for higher education. Some remained in England, principally in London, but others returned to Cyprus as doctors, nurses, pharmacists, attorneys, and civil servants. Other Cypriots were able to travel to what they considered their real home countries, Greece and Turkey, for the same purpose.

During the Cold War, Cyprus once again became a pawn in the battles between other powers: Greece and Turkey, the United States, Britain, and the Soviet Union. Activists in Greece, inspired by the notion that Cyprus should become Greek territory, supplied arms and funds to Greek Cypriots who were determined to oust the British from the island. The British director of medical services, P. W. Dill-Russell, made note of the insurrection in his annual report for 1955: "The inflammatory influences at work have given rise to acts of hooliganism . . . assassinations by fanatics and organized terrorism."[2] As the insurrection gained momentum, relations between the

Greek and Turkish communities deteriorated, since Turkish Cypriots had every reason to fear the consequences of union with Greece.

After five years of increasingly effective guerrilla warfare, British rule ended in early 1960. The negotiations that followed left the British with two fairly large naval bases on the island in perpetuity and the Cypriots with a very cumbersome bi-communal republican constitution. Neither the Turkish Cypriots (who had pleaded for partition) nor the Greek Cypriots (who desired annexation by Greece) were satisfied, and within a few years ethnic clashes began in earnest, culminating in a full-fledged civil war in the winter of 1963–64. Some of the fiercest battles of that short war occurred in the vicinity of the General Hospital in the capital, leading, among other things, to an almost complete breakdown in relations between the Greek and Turkish Cypriot medical professionals who had previously worked there in relative harmony.

A U.N. peacekeeping force was deployed to Cyprus early in 1964, but was unable to broker anything more than a hostile truce between the warring communities; violence still erupted periodically. Turkish Cypriot officials withdrew from the government, and many Turkish Cypriot families voluntarily enclaved themselves, seeking U.N. protection from their neighbors. The Greek Cypriots embargoed the enclaves—and for the next eleven years the Turkish Cypriots, subsidized by Turkey and the U.N., established the equivalent of a parallel governmental structure in the enclaves.

In the summer of 1974 a group of Greek and Greek Cypriot military men, assisted by the U.S. Central Intelligence Agency, attempted to depose the man who had been one of the leaders of the earlier insurrection and who had been president of Cyprus since 1960: Archbishop Makarios. Turkey used the occasion of the coup to mount an invasion, sending in approximately forty thousand troops. Within a month the Turkish army had occupied the northern third of the island, forcing the two hundred thousand Greek-speaking Cypriots who lived there to flee south. Although Makarios succeeded in returning from exile fairly quickly, the Greek Cypriot government was in disarray; under pressure from Turkey it agreed to what Turkish Cypriot leaders had wanted for the past twenty years: partition. By the spring of 1975, 40 percent of the Cypriot population—Greek and Turkish—were refugees, living in tent cities maintained, protected, and supplied by the U.N. High Commission on Refugees (UNHCR). The remaining Greek-speaking Cypriots in the north headed south, and Turkish-speaking Cypriots in the south headed north, though not always willingly. They were all leaving their property behind—orchards, fields, and homes that had been in families for generations—in a form of escrow that no one really trusted.

Today the refugee camps are long gone, but the Cypriot rift is only slowly being repaired, and the fate of the escrowed properties remains a matter of bitter contention. Two independent, not quite parallel governments exist on the island, separated by U.N. peacekeeping troops. In the north the Turkish Cypriots have an elected president and a legislature, and their government functions administratively as if their area were a semiautonomous region of Turkey, even though it declared itself an independent nation in 1985. Turkish troops patrol the borders, citizens travel on Turkish passports, and the Turkish lira is the official currency. The government does have independent taxing authority, and, with help from Turkey and the United States (funneled through the UNHCR), it has built a modern school system, with institutions of higher education, and a medical system that includes a large but poorly equipped public hospital. However, as it is unable to trade with any country other than Turkey (the only country that recognizes it as a nation), and as Turkey has its own economic troubles, Northern Cyprus has a very weak, undeveloped economy and has never been able to support itself without a substantial Turkish subsidy.

In the south, Greek Cypriots have an elected president and a legislature, which governs a truly independent nation that has its own armed forces, currency, and diplomatic relations with every other country in the world except Turkey. With a set of civil institutions in place, Greek Cypriots were able to recover from the civil disruption faster than their Turkish counterparts. Within a few years tourists returned in large numbers, and by the 1990s English and European entrepreneurs had discovered the tax advantages of using the Republic of Cyprus as an "offshore" location for their businesses. By the turn of the millennium, Greek Cyprus was a fully developed economy in which service workers far outnumbered agricultural workers, English was an almost universal second language, and postsecondary education was widespread. The Republic—without its northern section—was admitted as a member state of the European Union in 2004.

The capital city of Cyprus, Nicosia/Lefkoşa, remains one of the divided cities of the world. There is only one crossing point between the two halves; an old boulevard lined with the bombed-out shells of townhouses, land mines, and rubble. Foreign diplomats and U.N. officials continue to shuttle back and forth across the so-called Green Line in a valiant, but still not completely successful, effort at reconciliation.

When they departed in 1960, the British left the newly created Republic of Cyprus with an aging but adequate public health infrastructure. Following what was then the British model, in its very first years the Cypriot health system was a national health system; all facilities were maintained at government expense, treatment was available at no cost to patients, and everyone on the staff—doctors, nurses, pharmacists, technicians, janitors—was a government employee. As there were no specialist training facilities on the island, most of the medical professionals and paraprofessionals had been trained in Britain, Greece, or Turkey, although some, especially the technicians, had been trained on the job in Cyprus during a post–World War II antimalaria campaign sponsored by the Rockefeller Foundation.

The civil war of 1963–64 brought an end to this arrangement. The Turkish Cypriot decision to withdraw from the government also meant that Turkish Cypriot medical staff withdrew from government service. When the shooting stopped, Turkish Cypriot physicians and nurses began to construct a separate medical system independent of the old infrastructure. During the enclavement period (1964–1974) this medical system was underfunded and inadequate, but after the establishment of the Turkish Republic, the United States (through UNHCR) and Turkey contributed millions of dollars to build and outfit hospitals and clinics. Since the main British hospital in Nicosia and most of the smaller clinics in the port cities remained in the hands of Greek Cypriots, and since most of the Greek Cypriot medical professionals remained in the employ of the Republic, Greek Cypriots received better medical care than Turkish Cypriots during the decade of enclavement.

In the years since partition, a mixed medical economy has developed on both sides of the buffer zone. In the Turkish Republic of Northern Cyprus all medical care is free and all medical staff are employed by the government. Because of the weakness of the Turkish Cypriot economy, however, the public medical service is very poorly funded and medical staff members are very poorly paid. As a result, a private system, which is tolerated although not entirely permitted, has grown up alongside the public system. Most doctors and technicians work part of the day or part of the week in public service and moonlight in the private system for the remainder. Patients who can afford to pay for their medical care try to avoid the public facilities. Some patients also travel overseas for advanced medical care, sometimes to Britain (where they are entitled, as members of the Commonwealth, to subsidized care) and sometimes to Turkey.

In the much more prosperous Republic of Cyprus, public hospitals can offer better care and facilities. Some physicians, nurses, pharmacists, and technicians remain full-time employees of the ministry of health; they staff the hospitals and clinics on which the poor rely for medical care. Over the years, however, a private, fee-for-service medical system has grown up in the Republic, which offers, for those who can afford to pay, more spacious, modern, and efficient services, including some kinds of care that are not available at all in the public facility, such as *in vitro* fertilization and, more recently, pre-implantation diagnosis. Unlike their counterparts in the north, physicians and medical paraprofessionals in the private medical system in the south are not employed simultaneously in public service.

Despite their tragic history of civil unrest, the people of the two parts of Cyprus display many similarities: their culinary traditions, for example, are almost identical, and so is the architectural style of their nonreligious buildings. They also share a high frequency of the mutated genes that cause β-thalassemia. One in seven Cypriots, whether Turkish- or Greek-speaking, is a carrier of one of these mutated genes. Like the genes for sickle-cell disease and Tay-Sachs disease, the genes for β-thalassemia are recessive. The carrier rate for β-thalassemia on Cyprus is almost twice as high as that of sickle-cell anemia among African Americans and more than three times as high as the one for Tay-Sachs among Ashkenazi Jews. Before the screening programs started, on average, sixty Cypriot babies were born with this devastating disease every year.

β-thalassemia

What is now called β-thalassemia was once called Cooley's anemia, named for the Detroit pediatrician Thomas B. Cooley, who, with his pathologist colleague Pearl Lee, first described it in 1925.[3] In the clinic Cooley ran for poor children of Detroit, he had seen five

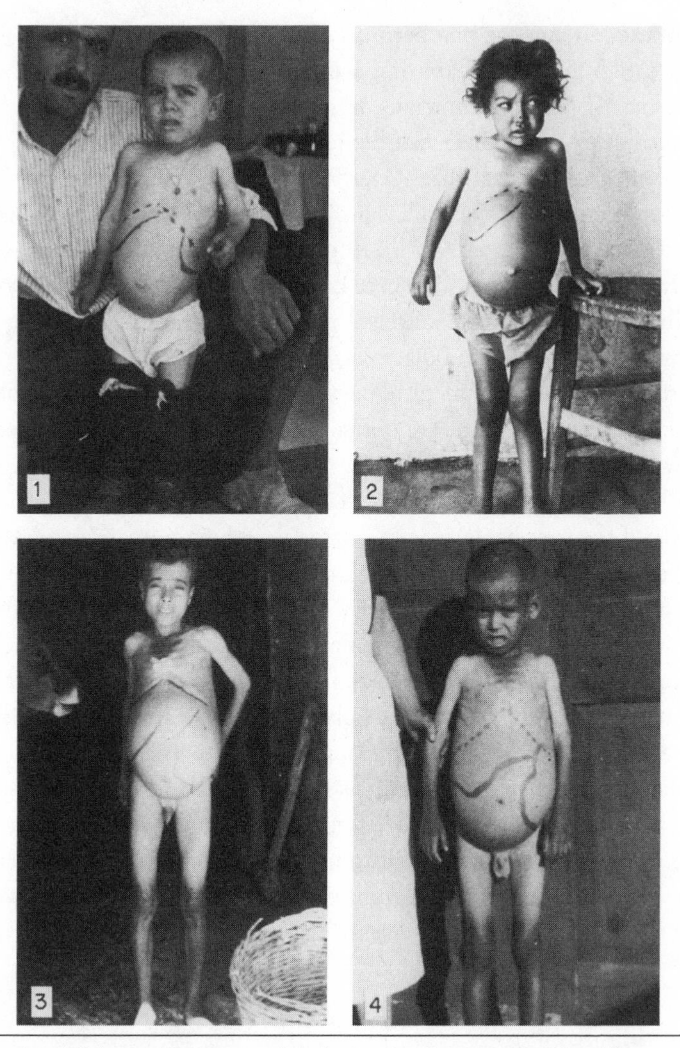

Cypriot children with untreated thalassemia, photographed by Allan Fawdry, M.D., in the early 1940s. The marks on the children's bodies are meant to highlight how large their spleens have become. The boys in (3) and (4) also show the characteristic deformation of the jaw and cheekbones.

From Bernadette Modell and Vasili Berdoukas, *The Clinical Approach to Thalassaemia* (1984), 268.

youngsters with an odd form of anemia: in addition to the usual symptoms (listless behavior, pallor, failure to grow normally), these children had enlarged livers and spleens, yellowish skin, and malformations of their facial bones which gave them what he called a "mongoloid appearance." Stumped as to the cause of this peculiar array of symptoms, Cooley turned to Lee, who soon discovered that the children's blood looked abnormal. There were only a few intact red blood cells, and those were unusually large and pale when stained for hemoglobin. When the children's blood was mixed with distilled water, the cells did not swell up and break apart as normal red blood cells would have done. Three of the five children did not respond to any of Cooley's attempts at treatment; they were dead by the time the paper reporting their case histories was published.

Over the course of the next decade this special, fatal form of anemia became established as a clinical entity: Cooley and Lee continued to report on more extensive investigations of additional patients, as did several other groups of physicians in American cities—Baltimore, Rochester, Boston—with large immigrant populations. By the mid-1930s these clinicians had more or less agreed that the disease was congenital (since symptoms began in infancy) and that it was, as Cooley and Lee put it in 1932, "not rare in children of Mediterranean ancestry."[4] The critical diagnostic signs were "failure to thrive," skin with a yellowish pallor, distorted facial bones, very enlarged spleens, and various abnormalities of the blood cells, especially osmotic resistance—the fact that the cells did not swell and explode in hypotonic solutions. The disease appeared to be untreatable; no matter what they tried—transfusions, iron pills, splenectomy (in the hope of stemming the tide of badly formed red blood cells)—most of their patients died before reaching their sixth birthdays.

Thomas Cooley was a prominent physician and was president of the American Academy of Pediatrics in the 1930s. Perhaps in deference to his prominence and his clinical acumen, many pediatricians and hematologists began referring to the newly described disease as Cooley's anemia. But Cooley, a modest person and a Quaker, preferred that his name not be used as the designator of a disease. His name for it was erythroblastic anemia, to signify that the red blood cells were apparently being destroyed. In 1932 George Whipple, a hematologist at the University of Rochester Medical School, came up with the name "thalassic anemia," which combines the Greek word for the sea with the generic disease category of anemia. In fairly short order, "thalassic anemia" became "thalassemia." (The Greek letter β was added two decades later, when biochemists discovered that, in this disease, one of the two chains of the hemoglobin molecule, the β-chain, does not form properly.)[5]

Cooley was opposed to eugenics and, perhaps for this reason, he was reluctant to say that the disease was either hereditary or racial.[6] Other physician-researchers, however, were convinced that the disease was somehow passed from parent to child. They had noticed, for example, that among their small number of thalassemic patients there were often siblings or cousins. Some physicians began constructing pedigrees and testing the blood of parents and (where possible) grandparents. The blood tests were revealing; although the parents of afflicted children appeared to be reasonably healthy, their red blood cells were also more resistant to hypotonic solution than normal.

A Greek physician, J. Caminopetros, may have been the first to provide evidence that β-thalassemia was a recessive condition in two papers he published in French in 1936 and 1938.[7] By 1945 Caminopetros's hypothesis had been corroborated by pedigree and blood studies done by several Italian and American researchers working in clinics with large numbers of patients. One of these American researchers was James Neel, who, with his colleague William Valentine, coined the names thalassemia minor for the heterozygotic condition and thalassemia major for the homozygotic, fatal form of the disease.[8]

Some physicians who were studying the genetics of thalassemia were simultaneously trying to stave off the death knell for their young patients. Several had tried blood transfusions and seen their listless patients perk up a bit after receiving the new blood. Following transfusion, as one of them said, "the anaemic child in a few hours is transformed from a helpless gasping creature into a more or less normal individual able to walk and play with pleasure." Unfortunately, the improvement, although real, was transitory: "In the course of a few weeks, he slips back rapidly into his original condition."[9] Yet because it appeared to be the best option available, in the late 1940s several physicians with thalassemic patients began to wonder if a regular regimen of transfusions, given whenever the child became listless, might not help.

Such a therapeutic regimen would have been exceedingly difficult before 1945. However, during World War II technologies for collecting, storing, and dispensing large quantities of blood—blood banking—had been developed, and after the war major hospitals in metropolitan areas were quick to adopt them. Consequently, during the 1950s and 1960s British and American pediatric hematologists were able to experiment with different transfusion routines: some transfused their patients whenever they became listless, while others tested blood hemoglobin levels regularly and transfused when levels fell below a certain mark.[10]

The best regimen for the children turned out to be keeping their hemoglobin closer to normal levels at all times by regular transfusions with high concentrations of red blood cells, starting as early as possible in the child's life, even as early as six months. In teaching hospitals in major cities this kind of difficult regimen—a massive burden for patients and families—had become standard therapy for thalassemic babies and children by the end of the 1960s. Close adherence to the regimen allowed children to grow and behave more or less normally: a long, often painful, day in the hospital once a month also meant suppression of the most obvious physical stigmata of β-thalassemia, so the children developed no protruding abdomen, no yellow pallor, no malformed facial bones.

That was the good news. The bad news was an unintended but easily predictable consequence of regular transfusions: damaging

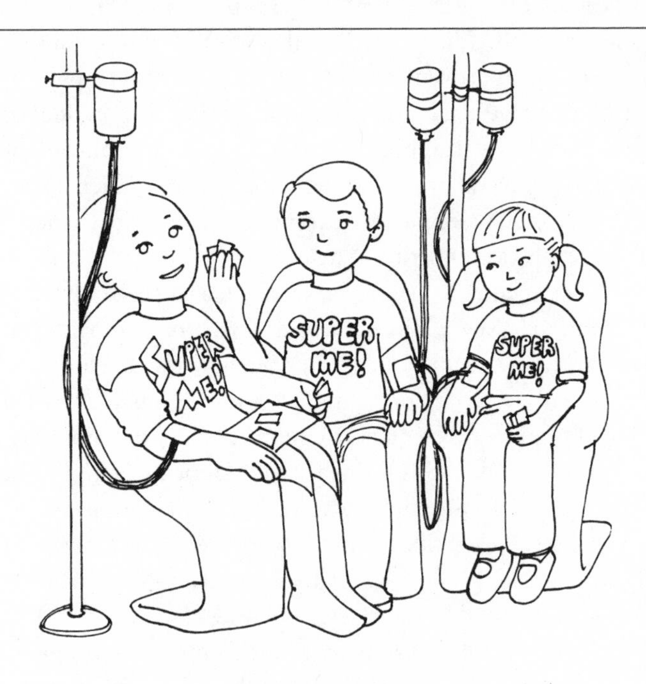

But I am lucky! To keep myself feeling good,
I go to the hospital where there are nurses and doctors to help me.
They answer my questions and give me medicine.
I also get blood packed with strong red blood cells!

Transfusion for thalassemia as represented in a coloring book designed to boost the self-esteem of thalassemic children.

From Sherry Bush, *My Coloring Book on Thalassemia: And How I Can Live a Super Healthy Life in My World* (1994), 13. With the permission of Sherry Bush.

amounts of iron began to accumulate in the children's tissues. As the first generation of regularly transfused thalassemic patients began dying in their teens—often because their heart muscles were struggling against iron overload—a few physicians began searching for a chemical that might somehow combine with all that iron to help the children excrete it.

The solution was an iron-chelating agent, deferoxamine mesylate, that had been put on the market in 1960 by the Swiss pharmaceutical company CIBA, under the trade name Desferal. This drug combines with elemental iron in the body to create a water-soluble compound that can be excreted. CIBA had intended Desferal to be used to treat other conditions in which iron accumulates in the body, but some physicians soon recognized that it might be useful for thalassemic patients on transfusion regimens. At first they tried injecting Desferal intramuscularly every time the patient received a transfusion. That worked, but only temporarily, because a day or so after the injection the patient's urine no longer contained any iron at all. By the late 1970s it became clear that the Desferal worked best if it was done slowly and continuously, by infusion under the skin rather than by injection. In order to avoid the ordeal of lengthy hospital stays for the infusions, physicians who were caring for thalassemic patients asked medical equipment makers to develop small, portable, battery-powered infusion pumps. By 1980 several dozen sets of parents in Britain and the United States had learned to mix Desferal powder with distilled water, stick a small hollow needle under their child's skin, and turn the pumps on for twelve-hour infusion sessions many times a week. The children had also, somehow, learned to tolerate this treatment. Desperation and love turned out to be powerful parental incentives; this was, the doctors had told the parents, the only hope for keeping their children alive past their teens.

Some parents gave up, but many others succeeded—and so did their children, when they became old enough to manage infusion themselves. Regular, often monthly, transfusions of concentrated red blood cells, coupled with regular, often daily, infusion of Desferal for ten to twelve hours at a time, has enabled thousands of β-thalassemia patients all over the world to live almost normal lives since 1980.

"Almost" normal is not, however, quite the same as "normal."

The needle is put on a box that works like a pump.
It runs on a battery. The box rests on my tummy or my bed when I sleep.
I think I'll give my pump a name.
I'll name it ____________________.

A coloring book page intended to help thalassemic children feel more comfortable with their dependence on infusion pumps.

From Sherry Bush, *My Coloring Book on Thalassemia: And How I Can Live a Super Healthy Life in My World* (1994), 18. With the permission of Sherry Bush.

Many patients who began treatment as babies and remained fairly compliant are physically and socially indistinguishable from their peers; they have gone through puberty (albeit with considerable help from endocrinologists), married, had children, pursued careers, engaged in sports—but they have done so at very great cost. The treatment regimen is time-consuming, painful, unremitting, and expensive. Compliance is very difficult; many patients fail for long or short periods of time, especially in adolescence, and end up compromising their health. Constant transfusion and infusion also bring many side effects in their wake: transient infections, hepatitis, allergic rashes, even, in some cases, AIDS. Desferal powder costs a great deal; the annual cost of infusion would be unaffordable, except for the wealthiest families, if it had to be paid out of pocket. Efforts to develop an effective oral chelator, which would obviate the need for infusion, have failed until very recently.[11]

Whatever the difficulties may have been, after about 1970 the diagnosis of β-thalassemia was no longer necessarily a death sentence for a child born where treatment was available. In the last decades of the twentieth century, wherever patients had access to a developed medical infrastructure, β-thalassemia changed from a fatal disease to a chronic one. Patients and carrier families, the people who suffer most from the disease, have repeatedly expressed their belief that the change has been for the better. Many parents of thalassemic children have become grandparents, a result which they could not have imagined when their children were young and which they regard, not surprisingly, as miraculous. There certainly are people who think that the therapy for β-thalassemia is as bad as the disease itself, but the relatives of β-thalassemia victims are not among them. In 1999, at a reception in honor of a β-thalassemia patient who had died at age thirty-five, a woman whispered to her companion, "So much pain. They should have just let him die as a baby." "Not if he was my child," her companion snapped in reply.[12]

β-thalassemia on Cyprus

Since the end of World War II, medical officials on Cyprus have been aware of the high carrier rates and, therefore, high incidence of β-thalassemia on the island. Allan Fawdry, a British physician posted to Cyprus in the 1930s, had developed an interest in the special form of anemia that he was seeing regularly in his clinics. Between 1940 and 1943 he examined twenty children who had enlarged spleens and deformed facial bones in addition to the usual symptoms of anemia. "Reference to my textbook of Anaemias," he wrote many years later, "showed that [they] suffered from Cooley's Anaemia, as the Americans called it."[13] There was no collection of medical journals on the island, so when Fawdry went on leave in 1943, he returned to Britain and read up on the disease. At that point he discovered the literature on Cooley's anemia trait, the condition found in the blood of the parents of thalassemic babies, which he described as "a peculiarity of the blood cells which, as it brings no harm to the possessor, can scarcely be classed as a disease."[14]

Having learned that a fairly simple test, the osmotic fragility test, would reveal possession of the trait, Fawdry decided to find out how frequent the trait was on the island. After examining the blood of five hundred Cypriot children, both Greek and Turkish, he concluded that about 17 percent of them had the trait (later called thalassemia minor or heterozygotic β-thalassemia), an estimate that has proved roughly accurate even with the use of more sophisticated diagnostic technologies.[15]

Fawdry was an evangelist, both for Christianity and for helping thalassemic children; one of his Cypriot colleagues described him as having "a stethoscope in one hand and a Bible in the other." His first concern was to distinguish β-thalassemia from curable diseases that have similar pediatric symptoms, such as malaria. "The certain diagnosis of the disease is of great importance," he wrote in 1946.

He added: "Though we cannot cure the disease we can at least save the parents the expense and the child the discomfort of being treated for diseases from which he is not suffering . . . [Some parents] reduce themselves to distraction and almost to destitution in visiting one doctor after another hoping against hope for a remedy."[16]

Fawdry's second concern was to ensure that every child identified as thalassemic received regular transfusions. The obstacles to this treatment regimen were numerous. Much of the population of Cyprus was rural, transportation was poor, and there was no blood bank. Many children, parents, and donors had to travel long distances to clinics, overnight stays were often required, and blood donors often asked for payments in lieu of lost wages. On top of all this, civil unrest was endemic, starting in the 1950s. Those who were familiar with the situation agree that many thalassemic children went entirely without transfusions or were inadequately transfused, especially during the early postwar years. A study of hospital admissions for β-thalassemia on Cyprus in the early 1970s concluded that some parents were still refusing treatment for their children and that only 18 percent of the known Greek Cypriot patients were receiving regular transfusions.[17]

In 1960, after Cyprus became independent, a Greek Cypriot physician, Minas Hadjiminas, became the director of pediatrics at Nicosia General Hospital. Over the next twenty years Hadjiminas and several of his colleagues became advocates for thalassemic children in Cyprus. They fought to hire special nurses and to create special facilities for thalassemic children. They encouraged parents to organize an association, the better to raise money for facilities and the better to pressure the government into helping meet their children's medical needs. Parents, physicians, and nurses traveled the length and breadth of the island lecturing about the importance of establishing blood banks and encouraging voluntary blood donations. Subsidies were sought from the World Health Organization, the British Council, and the U.N. High Commission on Refugees to help bring consultants to the island, send Cypriots abroad for specialized training, and purchase diagnostic equipment.

But the political turmoil in Cyprus between 1963 and 1975 meant, among many other things, that expensive medical services were not easy to deliver. Decades later Hadjiminas and his successor, Michael Angastiniotis, still spoke with bitter frustration about their inability to extend, or in some years even to maintain, adequate care for their Greek Cypriot patients.[18] For Turkish Cypriot patients the situation was even worse. Between 1960 and 1963 Nicosia Hospital had cared for Greek and Turkish Cypriot children equally, but after the winter of 1963–64, when the Turkish staff left government service and some Turkish families entered enclaves, their thalassemic children had to depend on the (sometimes surreptitious) humanitarian efforts of physicians on the two British naval bases.[19] During this period of turmoil, many families with thalassemic children, both Greek and Turkish, emigrated to Britain, where, as citizens of a Commonwealth country, their children would be treated without charge.

Accounts differ as to how and when the Cypriots learned about Desferal therapy. Some say that in the late 1960s émigré families whose children received chelation therapy in London informed their relatives in Cyprus; others say that one of the British physicians posted to the naval base, having learned of Desferal from a London colleague, began offering it to the children he was transfusing. Many people recall that, in the first few years, the children of affluent families had the easiest access to the drug, second only to the children who lived near to the naval bases, where the physicians were sometimes able to requisition it as part of their "necessary" supplies.

In the early 1970s a parents' association began to form among the Greek Cypriot families. The goal of the Cyprus Antianaemic Society was, according to its first bulletin, published in 1974, to pool resources and raise outside funds so as "to create, in Cyprus, the essential conditions to attain the highest possible percentage of survival."[20] Among those essential conditions were the establishment of better blood-banking facilities and the purchase of Desferal to dispense to patients whose families could not afford to buy it. Both goals were realized. Soon after partition, in the winter of 1974–75, the Antianaemic Society managed to convince the ministry of health to enlarge the existing blood banks, hire staff, and include Desferal supplies (and infusion pumps) in its budget, thereby ensuring that all Greek Cypriot children who needed transfusion and chelation therapies could have them, free of charge.[21] The church played a role as well, by deciding to build an entirely separate Thalassemia Center, with outpatient transfusion facilities and laboratories, adjacent to the Archbishop Makarios Hospital for Women and Children.

Turkish Cypriot families had a somewhat harder time of it until a stable, free thalassemia clinic was established in Lefkoşa in the late 1970s. Some Turkish children with β-thalassemia continued to be treated at the British naval base in the north, while others made the difficult decision to emigrate to London for treatment. Individual Turkish Cypriot families who could afford the expense purchased Desferal and pumps from the manufacturer and blood from Turkish soldiers stationed nearby.[22]

By 1980, when civil society had more or less stabilized on both sides of the buffer zone, virtually all of the approximately eight hundred children and adolescents with β-thalassemia on Cyprus were known to medical officials: two hundred or so of them in Turkish Cyprus, the rest in Greek Cyprus. In both countries, intense lobbying by physicians and parents had led to the creation of free treatment centers where children and teenagers could receive their transfusions and supplies of Desferal. The physicians, nurses, and technicians who staffed the clinics and the blood banks were all government employees; the blood itself was donated (in the north by Turkish soldiers, in the south by private citizens, encouraged by Antianaemic Society campaigns). Desferal was now being purchased, in bulk, at the expense of both governments, for distribution to patients at no charge.

Seven or eight years earlier, a WHO consultant had already warned at least the Greek Cypriot medical establishment that if it undertook to provide these new treatment regimens for all β-thalassemia patients it would soon face a medical/demographic crisis. George Stamatoyannopoulos, a medical geneticist at the University of Washington, arrived in Cyprus in 1972 with the intention of using biochemical tests he had developed to ascertain, as precisely as possible, the incidence and carrier rate of β-thalassemia on the island.

In a report to the ministry of health (then entirely in Greek Cypriot hands), Stamatoyannopoulos concluded that if the Cypriots were to offer the best available treatment regimens (transfusion with high concentration of red blood cells coupled with chelation therapy) to the current patient population, each patient would be likely to live into middle age—and perhaps beyond that. Medical costs would escalate, since each patient would need social support for his or her entire, now lengthier, life. To make matters worse, sixty to eighty thalassemic babies were being born every year, and they too were going to need social support for their entire, now lengthier life spans. Absent a prevention program, Stamatoyannopoulos estimated, the prevalence of β-thalassemia would change from 1:1000 in 1970 to 1:138 just fifty years later. Absent a prevention program, there would also be an increase of 300–400 percent in demand for blood and a rise of 600–700 percent in the cost of treatment. Cyprus would soon find that the needs of its thalassemic patients would completely engulf not just the available blood supplies but also the entire budget of the ministry of health.[23]

Cypriot cultural traditions did not lend themselves to β-thalassemia prevention, no matter how technically easy it was to identify carriers. Some carriers, the parents of children with β-thalassemia, were already well known, but physicians had had very limited success in convincing these people that subsequent pregnancies might produce more afflicted children. Worse still, physicians often could not convince carrier parents to allow testing of their healthy children or to take carrier status into account when arranging for the marriages of these other children. Even the adult siblings of afflicted children were reluctant to have themselves tested. In the 1960s Cyprus was still a very traditional, agricultural society; most marriages were arranged by parents. Most people were fatalistic about the results of their sexual relations; they also believed that a young person's value on the marriage market would plummet if anyone in his or her family was known to carry "the stigma."[24]

In the mid-1970s the physicians caring for Greek Cypriot β-thalassemia patients decided both to accelerate and to shift the focus of their prevention efforts. Henceforth they would direct their attention, not to the families of patients, but to young people who had not yet married. They lectured to groups of teachers, urging them (as advocates of modernity) to add β-thalassemia education to school curricula and to organize testing sessions for young women who were about to graduate. They wrote brochures to be handed out in schools, drafted scripts for radio broadcasts, and produced a documentary film. They obtained the consent of military officials to screen young soldiers who were serving their obligatory two-year army stint.

Although the public information campaign began in 1972 and a laboratory technician responsible for processing blood tests was added to the staff of the β-thalassemia center in 1973, in the first two years no dramatic results were evident.[25] "Many negative responses were noted amongst the young people," one physician wrote, "even those who had volunteered to be tested . . . Keeping the knowledge of heterozygote status a secret from family and friends was seen in many. Quite a few entered marriage without mentioning the matter."[26] Data on births in 1974 were not encouraging; there were 8,595 births in the Greek population, which meant that 54 thalassemic babies could have been expected. Fifty-one afflicted babies were actually born, three fewer than predicted—a result that could have simply been a statistical accident.

But it probably was not. Slowly, the number of people being tested began to rise and the number of thalassemic babies began to

Αρ. Ταυτότητας .... Αύξων Αρ. 213391

ΠΙΣΤΟΠΟΙΗΤΙΚΟ ΑΙΜΑΤΟΛΟΓΙΚΗΣ
ΕΞΕΤΑΣΕΩΣ

ΟΝΟΜΑ GOWAN Ruth

ΔΙΕΥΘΥΝΣΗ U.S.A

ΙΔΡΥΜΑ ΜΑΚΑΡΙΟΥ Γ΄
ΚΕΝΤΡΟ ΘΑΛΑΣΣΑΙΜΙΑΣ

ΓΝΩΜΑΤΕΥΣΗ

Η εργαστηριακή έρευνα έδειξε:

ΤΙΠΟΤΕ ΤΟ ΠΑΘΟΛΟΓΙΚΟ

Ο ιατρός

(Form Med. 232A.)

The identification card issued to the author when she had blood tests for thalassemia in Nicosia, Cyprus, in the spring of 1999. It does not reveal the test results, but it notes a ledger number at which the results can be found.

fall. The next year, 1975, the year in which 40 percent of the Cypriot population were displaced from their homes, the results were the same: three fewer thalassemic babies than predicted. In 1976, however, as civil life began to settle down, total births increased to 9,259, but only 37 thalassemic babies were born, significantly fewer than the expected 59. No one who was involved is able to

ΚΥΠΡΙΑΚΗ ΔΗΜΟΚΡΑΤΙΑ
ΥΠΟΥΡΓΕΙΟ ΥΓΕΙΑΣ

ΙΔΡΥΜΑ ΜΑΚΑΡΙΟΥ Γ'
ΚΕΝΤΡΟ ΘΑΛΑΣΣΑΙΜΙΑΣ

ΒΕΒΑΙΩΣΗ

Βεβαιούται ότι

ο/η...

από...

αρ. Ταυτότητας ...., έχει εξεταστεί για τη Μεσογειακή αναιμία και του/της δόθηκαν οι κατάλληλες συμβουλές.

Ο Ιατρός

Ημερομηνία............................

Το παρόν πρέπει να παρουσιάζεται κατά την έκδοση άδειας αρραβώνων ή γάμου.

This certificate, from the Thalassemia Center of the Republic of Cyprus, states that the person to whom it is issued was "examined for thalassemia and . . . was given counseling." The text at the bottom reads: "Give this certificate to the priest when applying for a marriage license." Like the identification card, the certificate does not reveal the diagnosis.

fully account for the magnitude of that change. Perhaps the message they were trying to convey had finally penetrated the fog of tradition, or perhaps the social disruptions of a coup, an invasion, and a resettlement program had emboldened young people to rebel against parental and religious strictures. Some women who already had a thalassemic child began asking the physicians in the clinics to help them obtain abortions, as the physicians were able to do, often at public expense, through contacts in Britain or Israel. Some married brothers and sisters of thalassemic children began coming in with their spouses to be tested. Some young couples who had decided to be tested before marrying had called off their weddings after discovering that they were both carriers. Others, upon learning the same news, asked for information about birth control, or to use the euphemism favored by the physicians, "refraining from reproduction."[27] Whatever the cause or causes, the data indicate that reproductive practices were changing, at least among young Greek-speaking adults on the island.

In 1977 the people in charge of the Greek Cypriot prevention program decided to shift gears once again—not because one year of encouraging data had met their goals, but because they had learned about a new technology of prevention: prenatal diagnosis of genetic blood diseases through fetoscopy. By 1974 at least two surgical teams had managed to insert a fiber-optic probe into the amniotic sac of a pregnant woman and to remove a small quantity of fetal blood from the umbilical vein for testing, without affecting the mother's health or the continuation of the pregnancy.[28] The diagnosis could not be made until at least the middle of the second trimester, when both the fetus and the umbilical cord had grown sufficiently large. This meant that an abortion following fetoscopy would be emotionally difficult for the pregnant woman, her partner, the physician, and the nurses.

Despite the novelty of the procedure and the difficulty of subsequent abortions, a team of London physicians who were working with thalassemic children decided to offer fetoscopy to their patients' mothers, explaining that there was some risk involved, but that the benefit of being able to bear an unafflicted child might be worth the risk. To the physicians' surprise, 95 percent of the pregnant Cypriot women to whom they explained the procedure requested it, and virtually all who were carrying a homozygotic fetus chose to terminate their pregnancies. "There was an overwhelming demand for this service from the British Cypriot community," they wrote, *"and a resultant reduction in the number of normal pregnancies terminated."*[29]

That last point—that prenatal diagnosis not only reduced the number of afflicted babies born but also reduced the number of abortions requested by carrier mothers—struck home in Greek Cyprus. Early in 1977 the physicians in the Greek thalassemia clinic began encouraging pregnant carriers to take advantage of this new procedure and began arranging for a government subsidy for travel expenses to Britain. Nine women went the first year; twenty-nine the next; eighty-nine the year after.[30] Not all were happy with their decision (one woman ran out of the examining room when she saw the probe and refused to return), but many were—and they voted with their feet, returning for fetoscopy not once, but two or even three times.[31] In those three years, 1977 to 1980, twenty-seven pregnancies of homozygotes were deliberately terminated (none were lost to miscarriage) and—equally important to the Cypriots—ninety-seven pregnancies were allowed to continue to term: ninety-seven healthy, nonthalassemic babies who, without prenatal diagnosis, might not have been born.

Prenatal diagnosis was an almost perfect fit with the pronatalist and antithalassemic impulses of the Greek Cypriot physicians in charge of prevention. They immediately decided to change their screening strategy from a focus on the unmarried young to a focus on pregnant women and the obstetricians who provided prenatal care. After a hesitant start, the new strategy began to succeed.

Minas Hadjiminas recalls that it took considerable effort to acquaint the island's obstetricians with the nature of β-thalassemia and the reasons for carrier screening, but that once educated, "most of the doctors fell in line."[32] In 1981, just four years after fetoscopy was introduced, he and Michael Angastiniotis were able to report (in Angastiniotis's words): "All ante-natal clinics, whether state or private, now demand a blood test for β-thalassemia on first visit."[33]

In 1981, 60 percent of the people being tested for carrier status were either pregnant women who were referred by their obstetricians or the spouses of pregnant women who had proven to be carriers. A year later, ten years after Stamatoyannopoulos had warned of a looming crisis, only eight thalassemic babies were born in the Greek portion of the island: 90 percent fewer than the seventy-three that demographers would have expected.

Mandating Genetic Screening

Prevention efforts took a somewhat different course in the Turkish Republic. During the period of enclavement, from the winter of 1964, when they left Cypriot government service, to the late 1970s, when they began to establish permanent medical facilities in Northern Cyprus, Turkish Cypriot physicians and nurses had all they could do provide basic medical care. Some Turkish Cypriot physicians were fully aware of the latest developments in β-thalassemia treatment and prevention (because they had children or close relatives with the disease) but were unable, during those years, to make them available (except to their own families, and then, sometimes, only by emigrating). As soon as a governmental apparatus had been established in Northern Cyprus, they sprang into action.

In February 1978 eighteen people, all of them parents of thalassemic children, most of them prominent members of the community, incorporated themselves as a national association to fight thalassemia. They quickly began to pressure the new government in the north to improve the treatment regimen for thalassemic patients. They raised money to help pay the expenses of several pregnant women who already had one child with the disease and who wanted to take advantage of prenatal diagnosis in London. Just as their Greek Cypriot counterparts had done a few years earlier, they started a public education campaign, giving lectures in schools, publishing brochures, and producing radio segments.

The Turkish Cypriot government responded to their pressure. At government expense, a pediatrician was sent to London for training. In 1979 Bernadette Modell, one of the British physicians who had pioneered Desferal treatment, was brought to Northern Cyprus to examine children and make individual recommendations for their treatment. Modell was also asked to document the need for a prevention program. (At the time there were roughly two hundred known patients and a prediction that between twenty and twenty-five babies would be added to the patient population every year.) Free screening facilities were in operation by 1979, but both the ministry of health and the members of the Thalassaemia Association were disappointed when only about a thousand young people volunteered to be screened.

So the Turkish Cypriots proceeded to do something that was, from a historical perspective, quite remarkable: they decided to coerce people into being screened. "We decided it was necessary to screen all the people who were about to get married," Ayten Berkalp, who was then the minister of health, explained a few years later. "So we amended the Turkish Family Law, by making pre-marital screening for Thalassaemia and genetic counselling compulsory."[34] The members of the Thalassaemia Association lobbied hard for passage. This was one of the first pieces of legislation passed by the newly elected parliament of Northern Cyprus, and the vote was unanimous.

The results of the newly mandated screening program were impressive. The law required that persons applying for a marriage license be screened for β-thalassemia trait and counseled by a specialist physician before the license would be granted. Because the law did not go into effect until the middle of 1980, only 2,800 people were screened that year, but the next year the figure rose to 5,056—and remained at that level for another decade, after which it began to fall as a result of the emigration of many young people prior to marriage.[35] At the same time, the Turkish Cypriots were continuing to take advantage of the facilities for prenatal diagnosis through fetoscopy: one woman went to London for the procedure in 1978; in 1980 ten, in 1983 forty-two. All of this, including the pregnancy terminations, was paid for by the Turkish Cypriot government.

In 1975, before any preventive measures had been taken, the Turkish Cypriots counted fifteen thalassemic newborns. In 1981, the first full year after premarital screening and counseling became mandatory, the figure was halved, to seven. By 1984, the year fetoscopy became available in Northern Cyprus itself, the figure had fallen yet again, to four. Between 1985 and 1996 only a dozen or so new cases were added to the patient register, most of which, in the early years of that decade, were the result either of misdiagnosis or of marriages that commenced before 1980.[36] Eight years after the Thalassaemia Association began its public education campaign, seven years after the first Turkish Cypriot woman traveled to London for prenatal diagnosis, and six years after the parliament mandated premarital genetic screening, virtually all young Turkish Cypriots had apparently accepted the wisdom of premarital screening, prenatal diagnosis, and abortion for homozygotic fetuses. They were voting, as it were, not so much with their feet as with their reproductive behavior.

Not long after the Turkish Cypriots mandated screening, the Greek Cypriots began to consider doing the same thing. By combining public education, carrier screening, birth control, abortion, and fetal diagnosis (the latter three at no charge to heterozygotic couples) they had brought the number of new β-thalassemia births down to 10 percent of expected levels, without having to mandate screening. Nonetheless, many concerned people felt that a reduction of 90 percent was not good enough, that a reduction of 100 percent, to zero β-thalassemia births, should be the goal. A voluntary system would never reach that goal, they argued; only coercion would work. The motives of those who favored mandated premarital screening were mixed. Some were still concerned with keeping the social cost of β-thalassemia treatment as low as possible; others wanted to reduce suffering as much as possible. "The suffering, the suffering; you can't imagine it," Minas Hadjiminas exclaimed many years later. "We had so many disasters. I saw one woman commit suicide. I saw kids left to die after the diagnosis . . . I wanted 100 percent; 100 percent was important to me. I don't want people miserable."[37] Public education efforts, meant to encourage people to volunteer for screening, were both time and energy consuming, not just for the medical staff but also for parent volunteers; mandatory premarital screening would allow them to attend to matters considered more interesting and valuable, such as research, therapy, and fundraising. The desire for a coercive system may also have resulted from a degree of ethnic competition with their neighbors to the north, the "anything they can do, we can do too" syndrome.

The Republic of Cyprus has an established church, which means, among many other things, that virtually all marriages are made legal by religious, not secular, authority. Mandated premarital screening therefore required the approval of the archbishop, not that of the legislature. In 1982 a small committee of prominent parents and physicians met with Archbishop Chrysostomos and the bishops who helped him make church policy. In laying out the case for a premarital certificate, the committee argued that the net result of premarital screening would be a steep reduction in abortions. A decade of experience with carrier screening had revealed that only a few engaged couples decided to end their engagements on the discovery that both were heterozygotes. In the early days, most had married, used birth control, and terminated all "accidental" pregnancies. With the advent of prenatal diagnosis, the committee continued, use of birth control had dropped among couples who already had one diseased child as well as among couples who knew, from screening, that they were both heterozygotes; the number of abortions had also dropped because these couples had stopped terminating all their pregnancies. Universal screening would mean that all heterozygotic couples would be alerted to the need to use prenatal diagnosis with all pregnancies, including their first. The net result would be, not only that the incidence of abortion would drop further, but also that hundreds of Orthodox Cypriot couples—some of them relatives and friends of the bishops in the room—would have children, since they would no longer be afraid to conceive.[38]

"We have a very pragmatic church," Minas Hadjiminas is fond of saying. Early in 1983 Archbishop Chrysostomos announced that, henceforth, priests of the Cypriot Orthodox Church could request that both partners to a marriage produce certificates, from the (Greek) Thalassemia Center, attesting that they had been tested and counseled. Two years earlier, when the Center's building, adjacent to Archbishop Makarios Hospital for Women and Children, was dedicated, Chrysostomos had used the occasion to deliver a long speech reminding everyone present that the church remained passionately opposed to abortion. Two years after the decision to mandate premarital certificates, the goal of zero β-thalassemia births was finally achieved.

Moral Congruence and Political Consensus

The history of mandated genetic screening on Cyprus has a great deal to teach us about the freighted question of whether compulsory genetic screening is right, wrong—or something in between. Whatever their other differences may be, the Turkish and Greek citizens of Cyprus seem to have agreed that their genetic screening policies are, if not wise, at least politically acceptable. They may be griping and worrying in private, but there has been no public effort either to challenge the law that mandates premarital screening in the north or to dissuade the Orthodox priests from requiring a premarital certificate in the south.

Two of the architects of the Greek Cypriot program, Minas Hadjiminas and Michael Angastiniotis, have offered several socioeconomic explanations for why mandated screening was acceptable to Greek Cypriots but might not be acceptable elsewhere.[39] Since the 1960s most Greek Cypriots have favored small families, and the total population is relatively small, literate, and homogeneous—all of which meant that a public education program had a high likelihood of success. In addition, because the population is small and the carrier rate fairly high (roughly one in forty-nine marriages is between heterozygotes), just about every Cypriot knows someone who has β-thalassemia. The fact that the population is genetically homogeneous also means that the risk of being a carrier is equally distributed through the entire population. In a different kind of population, where the risk might be very high for one social group and much lower for another, mandating screening for everyone—high risk or low—might be less likely to be either efficient or acceptable.

Socioeconomic explanations can take us only so far, however. The issues on which genetic screening programs touch—abortion, discrimination, caregiving, suffering—are all profoundly emotional and moral matters, and when people have very strong feelings about such matters, they will sometimes act against their own socioeconomic interests. In addition, the thalassemia screening programs are complex social institutions, requiring the cooperation of many different kinds of social actors. Complex social institutions cannot be created and maintained in democratic societies unless many different kinds of people, acting in several different social roles, somehow manage to reach a level of moral consensus, making political cooperation possible.

One group of crucial social actors in the history of Cypriot mandated screening was made up of government officials, legislators, and bureaucrats: the Turkish Cypriot legislators who voted in favor of mandating premarital screening, the officials of both health ministries, north and south, who approved the budgets that paid for so many aspects of the programs. Government officials were in favor of β-thalassemia prevention for reasons that were absolutely straightforward and ethically utilitarian: neither country could afford to provide the highest level of care to a growing population of thalassemics. The cost-benefit calculations had been made explicit in George Stamatoyannopoulos's report to the (Greek Cypriot) ministry of health in 1972: the population of thalassemics would double in fifty years; by 1990 half of all possible blood donors on the island would have to give blood annually just to supply the needs of thalassemics; the cost of treatment would soar from half a million to three million pounds a year, requiring the total annual budget for the ministry of health to double.[40]

By the early 1980s, when both governments started to cover the costs of high-level therapy and premarital screening (mandatory in the north, still voluntary in the south), the cost-effectiveness of prevention was obvious to anyone who bothered to look at the figures. In Greek Cyprus, each β-thalassemia patient cost £84,210 a year to treat (there were then about six hundred patients), but the entire prevention program cost only £130,696 to operate. More concretely: only two births had to be prevented per year in order for the benefits of prevention to exceed the cost.[41]

Some ethicists complain that cost-benefit analyses involving human lives are both immoral and inhumane, that, in other words, utilitarian ethics are not ethical.[42] Public health officials do not agree. They see themselves as being charged with ensuring the greatest degree of health and happiness for the greatest number of people *within the confines of the resources available.* Thus, in Cyprus, they have seen the decision to develop prevention programs—even mandated ones—for β-thalassemia as supremely ethical, enabling them to spread the limited monies at their disposal over a larger portion of the population, while at the same time covering the medical expenses of those who are already suffering with a chronic disease. Indeed, they see themselves as carrying forward the important ethical principle of public health and social medicine, namely of caring for all members of a population equally. Had they not supported prevention they could well have been charged with dereliction of their professional duties. Absent a prevention program, some Cypriot β-thalassemia patients were likely to have died for lack of funds to pay for the treatments they needed.

The Cypriot pediatricians who cared for thalassemic children were another crucial segment of the community that supported prevention. They regard themselves as a very special breed of physicians, since day in and day out, they spend their time with children who are suffering, and, day in and day out, they provide the kind of care that is dismissed as boring, unheroic, and medically unchallenging. They are motivated by the special passion they develop for their patients, especially the very young ones, and by a very intimate relationship with their patients' families. This mundane but intimate caregiving leads, they say, to a fierce desire to minimize the suffering of their patients and that of the parents with whom they must become partners in caregiving. When pediatricians who care for thalassemic children speak and write about their patients they tend to lose their tone of medical objectivity. Here, as just one example out of many, is what Allan Fawdry had to say about the experience of trying to give regular blood transfusions to a thalassemic child:

> When transfusion is repeated many times, severe and even fatal febrile reactions may occur; the technical difficulty of infusing blood

> into the minute veins of small children becomes well-nigh insuperable; the finding of compatible donors becomes almost as exhausting as the performance of the transfusion; *and one faces the metaphysical problem of whether for a child a continued life of semi-invalidism frequently punctuated by the unpleasant experience of transfusion is better than no life at all.*[43]

Thus, for physicians like Minas Hadjiminas, first director of the Thalassemia Center in Greek Cyprus, his colleague and successor Michael Angastiniotis, and Gülsen Bozkurt, first director of the Center in Turkish Cyprus, a β-thalassemia prevention program was, simply put, the best way they could envision *to reduce suffering,* to achieve a positive resolution of the metaphysical quandary that Fawdry had posed almost thirty years earlier.

These Cypriot physicians are not strong proponents of abortion; most would certainly not agree with the feminist argument that a pregnant woman has a right to control her own body and therefore a right to have an abortion whenever she wants one. On the other hand, when they weigh the suffering of a child with β-thalassemia and the suffering of the parents who are caring conscientiously for a child with β-thalassemia against the death of a homozygotic fetus, abortion seems to them to be the morally correct choice. As Michael Angastiniotis put it in the early 1980s, "The Medical attitude is that we are indeed taking life from a being with lack as far as we know of awareness, in order to prevent a painful life and a painful early death of a conscious being—often consciously and emotionally interacting with its environment."[44] For Angastiniotis, this conviction about the wisdom of abortion to prevent β-thalassemia is entirely consistent with his commitment to the principles of social medicine. "I still believe what I was taught as a medical student in England in the 1960s; preventive medicine is the best medicine to practice, because it is best for the whole society. Where β-thalassemia is concerned, screening *is* preventive medicine—and what it prevents is suffering."[45]

The parents, heterozygotes themselves, who were founders and supporters of the disease associations, also played a crucial role in the development of mandated screening. Just like that of the parents of Tay-Sachs babies, their advocacy was born of their suffering. Some, whose children had been born before regular transfusion was possible, suffered because their children were so terribly sick and then died so young—and because their other children might be socially damaged by the stigma of being potential carriers. Later, when treatment was possible but not easily accessible, some parents suffered when they could not afford treatment that they knew other children, whose parents were wealthier, were able to obtain. Subsequently, when effective treatment was available and accessible, all the parents suffered from the demands of that treatment itself: from having to restrain a toddler for hours of transfusion, from having to inject an infant with a sharp needle many times a week, from having to discipline a teenager to comply with life-saving but painful therapy. They also suffered when they had to deal with the reality of discrimination, when their children were treated unfairly by teachers, coaches, schoolmates, and employers.

Both parent associations, north and south, were founded with multiple goals. Parents wanted to undertake public education; they felt that too many of their compatriots were still ignorant about the nature of the disease and the possibility of saving children from an early death. They intended to lobby their governments to take over the expensive treatments so that even the poorest thalassemic child could receive them. They intended to raise money to augment facilities, hire staff, and subsidize research. And they also wanted, as the 1974 brochure on the Cyprus Antianaemic Society stated, to work for "the organization and application of a program for the detection of Thalassaemia carriers."[46]

They made good on this last intention, even though neither they (who already knew their carrier status) nor their children (who already had the disease) stood to benefit from it. The founders of the Turkish association actively lobbied for the law that made screening mandatory before marriage; the leaders of the Greek association arranged to present the case for a marriage certificate to the archbishop and his council. And the Thalassemia International Federation, which is headquartered in Nicosia, and which grew out of the Cyprus Antianaemic Society, today extends its services all over the Middle East and southern Asia, assisting other countries in establishing the kind of therapeutic and preventive programs that it helped to pioneer in Cyprus.

Reduction of suffering is one of the two moral rationales that parents give for their support of β-thalassemia prevention. They would like to create a world in which "no one else has to suffer the way my child and I have suffered." The urge to reduce other people's suffering is an urge that society rewards by designating it either as philanthropy or as altruism.

But something else was motivating the parents as well, something for which we have fewer positive descriptors. When asked to explain her advocacy of prevention, one mother said, "I did it to protect my daughter."[47] She wasn't referring to the effort to get the government to cover the costs of treatment, or the effort to establish a blood bank and raise the number of donors. Screening, she explained, was a way to protect her child by ensuring that the child received the highest possible level of medical care. Her family was not rich; they depended on free medical care for their daughter. But this mother knew that the money could stretch just so far; she feared that if additional patients were born, something crucial to her own daughter's care—blood, Desferal, hepatitis testing, *something*—would have to be rationed.

Some would say that this mother was being selfish, that she placed her own daughter's welfare ahead of the life of other people's unborn children. Others would disagree, arguing that she was being the opposite of selfish, that protecting a child's welfare is a moral priority—perhaps the highest priority—in the moral system of good parenting. When they joined physicians on the lecture circuit, when they donated funds to buy laboratory equipment, when they lobbied legislatures and church councils, when they stood on streetcorners distributing flyers, the parents saw themselves—and broadcast that vision to others—both as philanthropists and as good parents. Oddly, in English as well as many other languages, there is no single word or phrase to capture the moral essence of good parenting, nothing equivalent to "altruism" or "do unto others" or "thou shalt not steal." If it were ever articulated, however, one aspect of such a moral code would surely be precisely akin to the creed of politically active Cypriot parents of thalassemic children: "protect your children."

In a sense, the leaders of the Cypriot Orthodox Church were also acting as good parents, not to their own children exactly but to what they would not hesitate to call their own flock. They were protecting their flock in three related ways. First, by helping to prevent the birth of children who would suffer, they were reducing the overall level of suffering among their congregants. Second, they were acting to reduce the number of abortions sought by their congregants, since many couples had decided to remain childless (through contraception and abortion) rather than bring a thalassemic child into the world. Many people who regard abortion as an evil, as Orthodox clergy do, reasonably regard actions that reduce the number of abortions as minimizing evil. Third, by encouraging heterozygotic couples to reproduce, the church hierarchy was, also realistically, protecting its flock by helping to enlarge it. From the perspective of church officials, then, the reality of β-thalassemia created a moral calculus in which the wisdom of reducing suffering and protecting the welfare of living parishioners outweighed the wisdom of respecting the life-potential of embryos and fetuses.

Finally, two groups of Cypriots voted with their feet in favor of prevention through genetic screening: the pregnant women who subjected themselves to fetoscopy and the β-thalassemia patients themselves. The two groups were motivated, at least in part, by very similar moral concerns based on intimate experience with the disease and its therapies. Fetoscopy is an invasive procedure, and when it was first used to diagnose homozygotic β-thalassemia in fetuses no one precisely understood the risks of infection or internal bleeding for the mother, miscarriage, or debility for the fetus if carried to term. The Cypriot women who traveled abroad for the procedure were obligate heterozygotes; they had already given birth to a child with β-thalassemia. Their motivations were, no doubt, complex. Some may have been pressured into taking the risk by their partners or by members of the medical establishment. Some may have been acting selfishly, unwilling to take on the burden of caring for another gravely ill child. But even those who were coerced and those whose primary motives were selfish can also be understood to have been acting altruistically and in the interests of *both* their children; the one already alive and suffering as well as the one who might be born to suffer. Some of the physicians who performed the early fetoscopies acknowledged the brave cooperation of these women by thanking them, in print, by name, a most unusual practice.[48] Had these women known of the feminist critique of medicalized pregnancy or the disability-rights critique of genetic discrimination, they probably would still have put their bodies and their pregnancies at risk, in order to prevent the pain, suffering, and discrimination that they already had experienced firsthand as the mothers of children with β-thalassemia.

Many β-thalassemia patients have demonstrated their support for screening by participating actively in the advocacy groups that continue to flourish in both the Turkish and the Greek communities. Most patients also organize their marital plans around their belief in prevention, some by refusing to become romantically involved with other patients, some by refusing to marry a carrier, others by undergoing (or insisting that a spouse undergo) prenatal diagnosis. "I'm lucky," one engaged patient reported in an interview, "my fiancé doesn't have the gene. I wouldn't have agreed to marry him if he had had the gene."[49] In 1999, by which time many of the eight hundred or so patients on Cyprus who had begun transfusion and infusion therapy twenty or more years earlier had reached the age at which they were starting families, medical officials knew of only two sets of patients—one Greek, one Turkish—who had decided to marry. Thus most adult Cypriot β-thalassemia patients vote with their reproductive behavior in favor of the ethic of prevention espoused by both their parents and their physicians: no one else should have to suffer they way I have suffered.

Mandated genetic screening was politically acceptable to the Cypriots because they were able to see it as simultaneously morally right *and* anti-eugenic. Several different moral perspectives—utilitarian, medical, and parental—converged in justification of the policies: the utilitarian ethic of bureaucrats and advocates of socialized medicine; the medical ethic not just of physicians but also of religious leaders and parents favoring reduction of suffering; and the parental ethic that motivates mothers and fathers to protect the interests of their children and religious leaders to protect the interests of their congregants. This moral convergence was made public by all the parties to the consensus: they wrote articles and gave speeches, they advocated and lobbied, contributed time, effort, and money—and they conducted their own lives in accordance with their moral convictions.

The anti-eugenic features of the policy have been made equally public. Only screening is mandated in both Greek and Turkish Cyprus. Carrier marriages are not prevented, and although supported through education programs, financial subsidies, and reproductive counseling, neither prenatal diagnosis nor abortion is mandatory. Heterozygotic couples are free to reject fetal diagnosis and to refuse the option of abortion. The state obligates itself to care for every person afflicted with the disease, no matter how that person came to be born. Indeed, the government of Greek Cyprus had gone one step further by the spring of 1999; it had arranged matters so that heterozygotic couples who had remained childless because of their opposition to abortion could avail themselves of pre-implantation diagnosis (genetic testing of an embryo), free of any charge.[50]

Thus, in both Cypriot states, premarital genetic screening is simultaneously pronatalist and anti-eugenic. The message conveyed to carrier couples is the biblical injunction "Be fruitful and multiply!" rather than the eugenic injunction "You must not have children!" Francis Galton, Charles Davenport, and maybe even Hermann Muller would be appalled by a social program that encourages carriers of a deleterious mutation to have children.

Cyprus is a very small island, and neither of its two political entities commands much attention on the world stage. Nonetheless, by instituting a pronatalist, compulsory genetic screening policy, the Cypriots have shown the world that genetic screening can be both morally right and politically acceptable. What the Cypriots have done, to put the matter another way, is to purge eugenics of its negative, punitive, murderous, and discriminatory character and, simultaneously, to affirm its original meaning: beautiful heredity. With the right social arrangements, the Cypriots are telling the world, and with the right medical intervention, all of us, including those who happen to carry "bad genes," can, and should, pass beautiful heredity on to our children.

Conclusion

The Convention for the Protection of Human Rights and Dignity of the Human Being with Regard to the Application of Biology and Medicine was promulgated, with great solemnity, by all the member states of the European Union, on April 4, 1997.[1] The European Parliament had asked teams of doctors, scientists, philosophers, theologians, lawyers, bureaucrats, and politicians to come up with guidelines for implementing such new and troubling technologies as *in vitro* fertilization, genetic screening, and organ donation. These thoughtful people deliberated long and hard, but despite all their efforts, the agreement that they drafted, and that all the member states signed, is—at least at first glance—a very puzzling document.

It begins, as all such pronouncements do, with a preamble, reminding the member states of human rights declarations to which they had previously agreed. Then it proceeds, through fourteen chapters containing twenty-two articles, to outline all the conditions that must be either maintained or prohibited in order to avoid "the misuse of biology and medicine [which] may lead to acts endangering human dignity" or, to put the matter concretely, in order to avoid the ethical morass of Nazism. Chapter II, "Consent," reminds the member states that nothing can be done to a patient without the patient's freely given consent. Chapter III, "Private Life," asserts that each person has a right to keep medical information private, and also a right to be informed about information that a medical professional has gathered. Chapter IV, "Human Genome," prohibits discrimination on the basis of "genetic heritage," as well as germ-line genetic therapy and sex-selection. Additional chapters create rules and regulations for scientific research on human beings and for the removal of organs and tissues from living donors for the purposes of transplantation.

The last several chapters are concerned with legal matters, including how the Convention on Biology and Medicine can be amended, what to do if it is infringed upon, and how its provisions can be made to mesh with the laws of individual nations. At the very end, however, this strong-willed document suddenly waffles; it actually seems to take away all the rights it has just so carefully announced and defended. Article 26, "Restrictions on the exercise of the rights," is in Chapter IX, "Relations Between this Convention and Other Provisions," and it might well be called the "no, unless" provision.

> No restrictions shall be placed on the exercise of the rights and protective provisions contained in this Convention other than such as are prescribed by law and are necessary in a democratic society in the interest of public safety, for the prevention of crime, for the protection of public health or for the protection of the rights and freedoms of others.

No, Article 26 seems to say, you may not sterilize people involuntarily, as the Nazis and the Americans once did, *unless,* of course, you are a democratic society and you decide that the people you want to sterilize are likely to have children who will be criminals. Or, *no,* you may not create a genetic screening program that discriminates against people on the basis of their genetic heritage, *unless,* of course, you are a democratic society and your legislature can be convinced that the birth of people with a particular genetic disease is a threat to public health.

Doing Bioethics with Historians' Tools

The Convention on Biology and Medicine waffles, I believe, because many people, thoughtful people, are profoundly ambivalent about the matters it addresses. Religious leaders of many persuasions have inveighed against the use of *in vitro* fertilization, but thousands of people—devout people and secular people, right-wing people and left-wing people, committed feminists and antifeminists—have been voting in favor of it with their eggs, their semen, and their money for more than three decades. Organ donation is another topic on which thoughtful people cannot make up their minds. While some countries do not allow their citizens to ask for payment in return for kidneys, blood, liver segments, and the use of wombs, other countries do not seem to care. In some nations research on unwanted embryos (whether obtained through abortion or in preparation for *in vitro* fertilization) is absolutely prohibited, while in others it is permitted but regulated, and in still others it is simply ignored. Moral, political, and social arguments can be made, have been made, and are still being made in defense of or opposition to each one of these practices.

The various regimens that together constitute genetic screening—newborn screening, carrier screening, adult screening, prenatal diagnosis in all of its forms—also generate a great deal of ambivalence. On the one hand, there are the known facts about what the Nazis did in the name of eugenics. In article after article, the Convention expresses multinational repugnance about what happened in Germany in the years between 1933 and 1945. On the other hand, as those who struggled to compose the Convention knew perfectly well, in the last four decades of the twentieth century thousands of physicians screened millions of adults and tested millions of fetuses for genetic conditions of all sorts, ranging from the fatal (such as Tay-Sachs disease) to the frivolous (such as sex—as some people see it). Public health officials and legislators cooperate in this endeavor. Bureaucrats and legislators from the left, the right, and the middle, advocates of socialism and advocates of capitalism, elected and appointed policymakers all agreed and continue to agree to permit genetic screening and testing, or to look the other way.

How can this be? How can well-meaning people willingly cooperate in what seems, on the surface, to be a fascistic and racist enterprise? And how can well-informed people, fully cognizant of what was once done in the name of eugenics, agree to abide by a convention, a *covenant,* that seems to give governments leave to do it all over again? Bioethicists have tried to answer these questions philosophically, by attempting to generate ethical rules ("respect the autonomy of patients" is one such rule) that can create a moral foundation for understanding what was wrong with eugenics and what is right about genetic screening. Medical geneticists try to answer these questions scientifically, by explaining that the eugenicists had a weak evidentiary base for their policies, but that medical genetics has generated much more reliable knowledge on which to predicate action.

I set out to explore the questions historically, to parse out the connections between the past history of eugenics and the current practices of medical genetics. I decided, in other words, to do bioethical work with the tools I knew best how to use, historians' tools, searching carefully through the record of what had happened in the past, in order to find out not only why the Convention for the Protection of Human Rights and Dignity of the Human Being with Regard to the Application of Biology and Medicine waffles in the particular way that it does, but also why people of all religious and political persuasions continue to practice as medical geneticists and continue to avail themselves of the services that medical genetics offers.

What I found surprised me—and helped me put to rest my own decades-old ambivalence about having once permitted a physician to extract some of the amniotic fluid that surrounded a fetus that I did not hesitate to call "my baby." Part of that ambivalence arose, of course, from what might be called the existential reality of genetic testing; the fact that the results of the test may raise the specter of terminating an otherwise cherished pregnancy. I have now come to understand that another part of my ambivalence arose from my initially sympathetic reaction to the various arguments that many very vocal critics had been making and are still making today, against genetic testing. Their criticisms made sense to me, made me wonder whether I should not have just gotten up from that gurney in the fall of 1979 and refused to participate. After two decades of research and untold hours of thinking about the evidence I was accumulating, I have come to the conclusion that my sympathy with the critics was misplaced. As a result of my historical research, the ethical and political foundation on which the current practices of medical genetics have been built is one to which I now feel capable of giving my full assent, without guilt, without ambivalence, and without apology.

Critics of Genetic Testing

The politics of reproduction make some very strange bedfellows. Criticism of genetic screening arises from four different political quarters, which taken together form a very strange political alliance: left-wing intellectuals, reproductive feminists, disability rights activists, and opponents of legalized abortion, or as they are often called, pro-lifers.

Left-wing intellectuals object to biological determinism—the linked notions that biology is destiny, that our genes determine our fate, that in the development of our individual personalities and skills nature is much more important than nurture. These critics also believe that biological determinism is implicit in the practice of genetic testing. Their objections are simultaneously historical and political: historical because they see modern genetic testing as a continuation of the evil practices and discriminatory ideas of early-twentieth-century eugenicists; political because they view biological determinism as fostering conservative behavior, discouraging reform by blaming biology, not society, for a host of social problems.

Dorothy Nelkin and Laurence Tancredi explored and summarized these concerns in their book *Dangerous Diagnostics.* "Testing," they wrote, "can insulate organizations from change, increasing their rigidity and enhancing institutional control at the cost of individual rights." Schools, hospitals, insurance companies, governments, and businesses, they argued, are all powerful social institutions that create and sanction genetic tests and that could use them to enforce conformity to their own norms of operation. Carrier testing might be used to constrain individual freedom in order to prevent the birth of children whose medical care would be costly. Prenatal diagnosis for Down syndrome could be sanctioned by governments so that schools would not have to adjust to educating "difficult" children. The end result, they worried, would be a society like the one depicted in Aldous Huxley's *Brave New World,* a society that abhors change, ostracizes rebels, and breeds people to fit fixed occupational roles. Abby Lippman raised the same set of concerns in her oft-quoted essay on what she calls "geneticization"—the notion that all human characteristics, including but not limited to diseases and disabilities, are fundamentally genetic. "Geneticization," she wrote, "articulates certain values and perspectives . . . [and] it would be naïve to think that how and for whom it is promoted and applied will not reflect and, in turn, reinforce these attitudes."[2]

In addition to being on the political left, Lippman is a reproductive feminist—a member of a group that separated from the feminist mainstream in the 1980s because of their opposition to the new reproductive technologies, particularly *in vitro* fertilization, surrogate motherhood, and prenatal diagnosis. These feminists created an organization to express their concerns, FINRRAGE (Feminist

International Network for Resistance to Reproductive and Genetic Engineering), and began to publish the *Journal of Reproductive and Genetic Engineering*. Both the organization and the journal are now defunct, but the set of attitudes they propounded has had a lasting impact, both on academic feminism and on the behavior (or, more tellingly, on the guilt feelings) of many feminists.

Lippman and other reproductive feminists did not approve of what they took to be the eugenic implications of the new reproductive technologies. They thought that women should not be seeking "perfect" babies; that the desire to have a baby genetically linked to oneself was a holdover from a eugenic and patriarchal past; and that the linked desire to have a genetically "healthy" baby was a fantasy perpetuated by a scientific and medical establishment permeated by eugenic ideals. Like the left-wing intellectuals, reproductive feminists also did not like the biological determinism that they believed was implicit in all the new technologies but most particularly in genetic testing, especially the notion that nature (that is, the genes) was more significant in human development than nurture (that is, mothering).

Reproductive feminists condemned genetic screening for other reasons as well, reasons that went beyond the left-wing complaints about eugenic and deterministic thinking. Some of those complaints were summed up by Patricia Spallone in her book *Beyond Conception:* "The growing repertoire of genetic technologies places more physical and psychological burdens on women, more reasons to undergo high-tech, scientist controlled means of reproducing." "The application of genetic technology," Spallone concluded, "is a further medicalisation of women's reproduction."[3] Reproductive feminists disliked all aspects of the medicalization of reproduction and pregnancy: they didn't like childbirth anesthesia any more than they liked amniocentesis, or bottle-feeding any more than, say, preimplantation diagnosis. All medical interventions seemed to them to be methods by which men gained greater control over women's bodies. They complained that under the "medical gaze," natural processes—intercourse, pregnancy, childbirth—are turned into pathological ones, which need to be managed by recourse to technologies and technological expertise, leading women to devalue their own bodies and their own understanding of their bodies and their desires. They didn't even approve of female obstetricians and gynecologists, regarding them as women who had been co-opted by the patriarchal establishment.

Reproductive feminists also criticized genetic testing as being tantamount to the industrialization and proletarianization of mothers. "We are looking at motherhood," Barbara Katz Rothman said, "as a production process, as cheap labor, as work . . . Genetic counseling is serving the purpose of quality control and wrongful life suits [malpractice suits brought by American women against their physicians for failure to offer genetic testing] are a variety of product liability legislation." Other reproductive feminists worried that since genetic testing was expensive, genetic diseases and disabilities would be eliminated from the middle-class population and would accumulate in the poor. Yet others attacked the development of genetic testing as unwarranted experimentation on women's bodies. On top of all this, reproductive feminists were very aware that several forms of prenatal diagnosis were being used, particularly in South and East Asia, for sex-selection of male fetuses, a practice Gena Corea designated as "gynicide."[4]

One last criticism raised by reproductive feminists connects them both to the disability rights movement and, oddly, to the profoundly antifeminist movement to reverse the legalization of abortion. Patricia Spallone put her finger on it when, compiling a list of complaints about genetic screening, she added that it creates "more reasons to contemplate selective abortion."[5] Reproductive feminists, in short, disliked genetic testing precisely because it leads to abortion for fetal indications. Spallone (and many other reproductive feminists) never used that phrase, however; they preferred "selective," no doubt because it resonates with the "selection" process that the Nazis used to separate those concentration camp inmates who were chosen for the gas chambers from those chosen for work details.

Disability rights activists use the same terminology. Adrienne Asch, one of the leaders of the disability rights movement, speaks of "prenatal diagnosis and selection of embryos" and of "selective abortion."[6] Marsha Saxton, another leader of the movement, entitled one important article "Disability Rights and Selective Abortion."[7] Like left-wing intellectuals and reproductive feminists, disability rights activists have a number of reasons for disliking prenatal diagnosis. They see it as discriminatory because negative attitudes about the value of people with disabilities are built into the system. They also see it as discriminatory because it creates false stereotypes about disabled people, suggesting that all disabled people are similarly unable to lead full, happy, productive lives. As Asch puts it, "My moral opposition to prenatal testing and selective abortion flows from the conviction that life with disability is worthwhile and the belief that a just society must appreciate and nurture the lives of all people, whatever the endowments they receive in the natural lottery."[8]

Like left-wing intellectuals, these activists are convinced that genetic testing depends on biological determinism and that biological determinism necessarily leads to social passivity, to the conviction that social reform can never ameliorate the plight of those who have "lost the toss of the biological dice." Disability rights advocates do not think—as they believe medical geneticists do—that disabilities are "diseases," or that disabled people are somehow "defective"; indeed, they would like to discard all negative notions about disability by discarding the word "disabled" in favor of the more positive term "differently abled." They worry that medical genetics will succeed in preventing the birth of differently abled people and that, as a result, the services available to such people will decline with their numbers in the population. Finally, the fact that genetic testing leads parents to terminate the pregnancies of fetuses that might be born blind or deaf or dwarfed or mentally retarded or lacking limbs fills disability rights activists with horror because of what they think it portends for the treatment of children and adults with these conditions. The choice to abort a fetus who will develop into a differently abled person is not a choice, these activists believe, that moral people should be making.

This, of course, is the basis for their alignment with pro-lifers, the people who are committed to ending the legalization of abortion. Adamant pro-lifers believe there are no indications—no fetal indications or social indications or psychiatric indications—that justify abortion, except, perhaps, when continuation of a pregnancy threatens the life of the mother. Pro-lifers believe that human life begins at conception, that the biblical commandment "Thou shalt not kill" extends to embryos and fetuses, and that (in the United States) the constitutional guarantee of a "right to life" should also extend to embryos and fetuses. Genetic testing followed by parental decisions to abort certain fetuses, pro-lifers argue, implies that some lives are not worth living, that some people are less valuable than others, and that the parents, physicians, and scientists who participate in such abortions deem themselves worthy of abrogating God's judgment.

Many people who are opposed to abortion are religious fundamentalists and, as such, they equate the evils of abortion with what they see as the evils of scientific materialism in general and Darwinism in particular. Many, many generations of biologists have regarded science in general and Darwinism in particular as opposed to religion in general and revealed religion in particular—and many, many advocates of revealed religion have returned the compliment by referring to biology, evolutionary theory, and genetics as prime examples of scientific materialism.

In other words, religious fundamentalists criticize genetic screening partly because it involves abortion and partly because it seems to them to be a product of scientific materialism. If matter is the only thing that exists, the argument goes, there is no God and there is no God-given soul. Without God or a soul, all people think of themselves as interchangeable and robotic. If people are interchangeable, then so are embryos and fetuses; if people are interchangeable, then no one is unique, and uniquely valuable, and neither is a fetus. Raising a disabled child is, from this perspective, an act of personal moral courage, and, in the words of the anti-abortion blogger Bill Muehlenberg, these children bring "great love and joy . . . and wonderful lessons learned and character development that takes place as a result of allowing these precious children to live." In addition, from the perspective of some religious fundamentalists, people who are willing to abort, or to perform an abortion, or to advocate legalizing abortion, must be narcissistic and greedy (another, linked meaning of materialistic), selfishly interested only in what is good for them, not what is good for their unborn child, their community, or their society. "We live in a throw-away society, and if we can't get the perfect baby, simply abort and try again," as Muehlenberg has put it.[9]

Like all the other critics of medical genetics, opponents of abortion consider it a contemporary incarnation of the eugenics movement. Muehlenberg refers to medical genetics as "the revival of eugenics as normative social policy in the West." He defines eugenics, just as left-wing intellectuals, reproductive feminists, and disability rights activists do, as "the search for a good birth, the search for a perfect baby, the search for a culture free of 'abnormal' people." Reprising a common theme, he refers to women who refuse to participate in genetic screening or who carry fetuses to term against the advice of their doctors as having "resisted the ideology of quality control and the paradigm of perfection." And finally, he appeals, just as all the others do, to the ultimate negative historical memory.

"Of course the last time we saw eugenics unleashed on a society as a whole was in Germany in the 30s and 40s. Never again, we said thereafter. But we have short collective memories."

Letting the Evidence Speak

Despite its power and pervasiveness, the connection that all these critics make between medical genetics and eugenics is historically fallacious. Activists on the political right are as mistaken as activists on the political left: genetic screening was not eugenics in the past, is not eugenics in the present, and, unless its technological systems become radically transformed, will not be eugenics in the future. The technologies of medical genetics were not constructed with eugenic goals, and the practices of medical genetics will not produce eugenic results—neither directly through the actions of medical experts nor indirectly, as the sociologist Troy Duster likes to say, through the "backdoor" of patients' compliance with experts' instructions.[10]

There is, to start with, no meaningful historical connection between the enterprise once called eugenics and the enterprise now called medical genetics. There were certainly some Mendelian geneticists, like Davenport and Muller, who were eugenicists, but the vast majority of classical geneticists figured out, early on, that most of the eugenic claims made about the inheritance of feeblemindedness and alcoholism were, if not entirely false, then at least undemonstrated.[11] Human genetics, the genetics of probability and statistics in large populations, the genetics of allelic frequencies and mutation incidences, indeed had some eugenicists among its founding fathers and mothers, but medical genetics owes very little to it. Medical genetics is, first and foremost, a medical specialty. During the formative years of the specialty, roughly between 1960 and 1980, most physicians were not interested enough in human genetics to take the trouble to understand it, largely because human genetics was then focused on evolutionary questions, not on matters pertinent to clinical practice.

The foundational science for medical genetics is classical Mendelian genetics: the genetics of dominant and recessive genes strung like beads along chromosomes. This is the genetics that provided the foundation on which the technological systems of genetic screening—newborn screening, carrier screening, prenatal diagnosis—were built. Classical geneticists were not, by and large, eugenicists, and neither were the physicians who constructed those systems on the foundation that the geneticists laid. By the time molecular genetics, the genetics of DNA and RNA, of sequences of molecules and production of proteins, had advanced far enough to be useful for screening purposes (the mutation that produces sickling hemoglobin was the first mutation whose location was pinpointed, in 1978) the technologies of genetic screening had been routinized for more than a decade.[12] Some of the scientists who lobbied for funding of the Human Genome Initiative may have used the rhetoric of eugenics when speaking to venture capitalists, legislators, and bureaucrats, but most of them (for example, Robert Sinsheimer, Joshua Lederberg, and James Watson) were molecular biologists, not medical geneticists; they had no medical training or clinical experience, and they played no role whatsoever in the development of the technological systems through which genetic screening is practiced.

Technological systems are built to achieve certain goals; those goals get hard-wired, as it were, into the components of the system. The chief goal of the eugenicists, "improvement of the race," was never one of the goals of genetic screening—and it did not become one, even after genomic research had identified the locations of dozens of disease-causing mutations. The founders of eugenics differed about which race they had in mind: some meant "the white race," some "the German race," some the "Mexican race" and some, even, "the human race." The founders of medical genetics, however, made deliberate efforts to separate themselves from what James Neel called "the parlous intellectual state" of eugenic research and practice because they thought it politically and scientifically correct to do so.[13]

From the very beginning, the founders of medical genetics—people like Neel, Fritz Fuchs, Michael Kaback, and Robert Guthrie—viewed their basic project as the relief of human suffering, not improvement of the race. Relief of suffering might, in their view, also improve the health of races or populations or societies (as, for example, it has done on Cyprus), but improving the health and wellbeing of individuals was always their primary goal. Some of the individuals they wanted to help were people who had already borne children who were struggling with painful and disabling diseases like β-thalassemia or muscular dystrophy. Geneticists wanted to reduce the suffering of both parents and children by helping the parents to have additional children, and by ensuring that those additional children would be free of the disease. Other individuals who could be helped were those who knew themselves to be at risk and were repressing their desire to have children for fear that the risk would become reality, that the risk status of the parent would be visited upon the child.

The reproductive goals of medical genetics are thus precisely the opposite of those of eugenics. Eugenicists wanted to ensure that the people they defined as genetically unfit did not reproduce; that is why they pushed for sterilization and segregation. Genetic screening was developed by medical geneticists to help the genetically "unfit," precisely the people the eugenicists would have sterilized, have as many children as they wanted. The earliest patients who were referred to medical geneticists for counseling were people who suspected (or whose doctors suspected) that they carried the genes for serious diseases. Prenatal diagnosis was developed to assure those worried patients that they could have children free of the disease they feared. To put the matter another way: the practices of genetic screening are inherently pronatalist. By supporting parental hopes for reasonably healthy children, they encourage at-risk couples to reproduce. As long as the regimens of newborn, carrier, and prenatal screening are not altered, medical genetics cannot produce eugenic results.

The vast majority of eugenicists, including the Lamarckian eugenicists, were progressives who wanted the state to take more control over the social and economic lives of its citizens. Mandated sterilization, eugenic courts, public institutions that segregated the feebleminded by sex, government subsidies for large families: these were all advocated on the grounds that the state needed to protect the health or the size or the quality of the population. Medical geneticists have been extremely reluctant to advocate state mandates because they are aware of what was once done by governments in the name of eugenics, and because they are aware of how the bungled state mandates for sickle-cell screening helped to destroy that public health initiative. Since the end of World War II medical geneticists have advocated mandated screening in only two situations: when identification of homozygotes (for PKU, for example, or sickle-cell disease) will make it possible to provide early and effective interventions; and when state-funded testing and abortion will simultaneously reduce individual suffering and prevent an insuperable financial burden to a community (as in the Cypriot screening programs for β-thalassemia). Their reluctance to advocate state mandates extends into their advocacy of legalized abortion, most particularly for fetal indications. Put in terms that will be familiar to feminists, such advocacy amounts to saying that parents, not the state, must be able to choose which children they will bring into their families.

The presumed connection between eugenics and medical genetics is not the only criticism of genetic screening that turns out to be historically ill-founded; there are several others. Reproductive feminists claim, for example, that genetic screening techniques were developed by inappropriate experimentation on women's bodies—but it seems unlikely that Fritz Fuchs's first amniocentesis patient, who was able to carry her pregnancy to term without fearing that she would give birth to a boy with hemophilia, or the several dozen Cypriot women who subjected themselves to fetoscopy to ensure that they would not raise another child with thalassemia, would agree.

Disability activists claim that genetic screening is a form of discrimination against the disabled—but it seems unlikely that the parents who banded together to form associations like the National Tay-Sachs Disease Association or the Cyprus AntiAnaemic Society or the National Association of Retarded Citizens would agree. The Tay-Sachs parents hoped for a cure for the disease that afflicted their children, and when a cure proved elusive they turned to prevention as a second-best—but still worthy—undertaking. The Cypriot parents sought both to lessen discrimination against their children and to prevent other afflicted children from being born, justifying both goals with the moral imperative to protect their children (and their grandchildren). The American parents who, through NARC, supported the campaign for mandated PKU screening also did not appear to regard prevention of one form of mental retardation as implying that people who suffered with other forms should not be treated as valued and valuable individuals. As African-American communities in the 1970s did not speak with one voice about the wisdom of screening for sickle-cell disease, so too the communities of disability activists do not seem to speak with one voice about the wisdom of screening for Down syndrome or any number of other detectable disabling conditions: some parents of afflicted children, including many who are politically active in their children's interests, hope to have more children (and grandchildren) but only if they can be assured that those children will not be similarly afflicted.

Disability activists also worry that as genetic screening diminishes the number of children born with disabilities, services to disabled people will decrease. There is no evidence to suggest that this has happened, even though screening programs have existed for decades. In Cyprus, discrimination against thalassemic people has lessened as the number of babies born with the disease has fallen. In the United States, the institutions to which parents once sent disabled children they could not care for at home have been closed, school systems have been required to provide special educational services for such children, and all kinds of public facilities have been modified to accommodate their needs—in the very same years that prenatal diagnosis became common practice. In most states, legislation has made it illegal for insurance companies to drop clients who decide to go forward with pregnancies after receiving worrying prenatal diagnoses. Contrary to what some disability activists assume, many people support widening the rights of disabled children and adults while simultaneously believing that abortion for fetal indications is morally wise.

Reproductive feminists have also claimed that what they call selective abortion violates the essential morality of motherhood, turning the ideal of nurturing care into the ideal of quality control. The legions of feminists (as well as the legions of women who are not feminists) who have sought—and even sued to get—prenatal diagnosis for Down syndrome are not likely to agree with them. Such women are, indeed, rejecting some traditional ideas about motherhood, particularly the notion that the ideal mother is willing to sacrifice herself for her children. By choosing prenatal diagnosis and the possibility of abortion, such women are asserting that they would prefer not to sacrifice themselves to caring for a chronically dependent, suffering child, and that there are other significant social roles that they hope to play in the course of their lives—spouse, employee, friend, athlete—that would be very difficult or impossible to combine with such care. Many of these women have put off childbearing to complete their educations or build employment resumes, which means that they are at increased risking of bearing a child with Down syndrome. "We have been oversold on the health advantages of starting our families early," declared Barbara Seaman, a well-known feminist medical writer, in the pages of *Ms. Magazine* in the winter of 1976.[14] By using amniocentesis, she went on to argue, women could now pursue education, career building, and childbearing—in the popular terminology, they could hope to "have it all"—no matter how long the first two may have delayed the third. Since they regard termination of pregnancy as emotionally difficult but not morally reprehensible, the kinds of women Seaman was addressing come to see prenatal diagnosis—and abortion for fetal indications—as part and parcel of their liberation, not, as the reproductive feminists would have it, as a source of oppression.

Reproductive feminists and abortion opponents have also argued against genetic screening on the grounds that it increases the frequency of abortion—but it is unlikely that women (and men, for that matter) who know they are carriers of single-gene mutations will see it quite the same way. Were it not for prenatal diagnosis, these people might terminate *all* the pregnancies they carried or caused to be carried. The physicians who created thalassemia screening programs on Cyprus would not agree either; they knew how often the parents of their patients were terminating pregnancies. The members of the council of the Cypriot Orthodox Church in the early 1980s might not agree either, since reduction in the frequency of abortions was one of the grounds on which physicians and parents managed to convince the church that mandated screening would be wise.

The technologies of genetic screening reduce, rather than increase, the likelihood of selective abortion. Fritz Fuchs and Povl Riis understood this when they explained, in 1960, that absent the ability to find out whether she was carrying a boy or a girl, their first patient, a known carrier of hemophilia, would have terminated her pregnancy—and that the committee charged with giving her permission would have approved her decision. Prior to the advent of carrier screening and prenatal diagnosis, women who knew that their offspring were at risk for a serious disease tried not to become pregnant, and if they did become pregnant they sought abortions. After the advent of prenatal diagnosis either 50 percent or 75 percent of those potentially afflicted pregnancies were likely to be carried to term: a significant reduction in the frequency of abortion. One abortion performed would be outweighed by two or three that did not happen.

In short, in the population of at-risk couples, genetic screening reduces abortions, or as M. Neil MacIntyre, an early proponent of prenatal diagnosis, put it, "The most spectacular and important aspect of . . . genetic diagnosis and management is that it preserves the lives of normal babies that would otherwise have been aborted because of a known genetic risk." MacIntyre's analysis is worth paraphrasing in its entirety, since it may help us to understand why some opponents of abortion—for example, priests of the Cypriot Orthodox Church—are willing to accept abortion for serious fetal indications. MacIntyre argued that prenatal diagnosis offered hope, that it was life-producing, life-enriching, and life-enhancing. "Prenatal evaluation and selective abortion," he wrote, "can enable parents who otherwise would not undertake a pregnancy to do so." It is also life-enriching to parents, he argued, as "knowledge of the existence of a genetic risk is almost certain to engender some fear or concern regarding pregnancy. The resulting anxiety is potentially a serious threat to the stability of a couple's sexual relationship, and some degree of deterioration of this aspect of their marriage is almost inevitable." Finally, the possibility of prenatal diagnosis is also life-enriching to siblings of an afflicted child and to other members of the extended family, MacIntyre said, because it can "free individuals closely associated with a genetically based tragedy from the suspicion and condemnation they have confronted in the past and allow them to live far more pleasant and rewarding lives."[15]

MacIntyre's approach is shared by the community of medical geneticists and genetic counselors, the people who spend their professional lives working with patients who are at risk. His approach suggests not only that the critics of genetic screening need to reassess their objections but also that the practices of genetic screening rest on a complex moral foundation, constructed from the medical injunction to reduce suffering combined with the parental injunction to protect children from harm.

The Convention for the Protection of Human Rights and Dignity of the Human Being with Regard to the Application of Biology and Medicine waffles because all the criticisms of genetic screening—especially the false analogy between medical genetics and eugenics—have been culturally powerful and pervasive. They are powerful because the Holocaust is a traumatic icon of evil; they are pervasive by virtue of having been voiced by ideologues of both the political right and the political left.

Ordinary people who have been swept up in the regimens of genetic screening without quite knowing what they were doing, or why they were doing it, have been as ambivalent as the experts who composed the Convention, and for the same set of reasons. Antieugenic rhetoric is good political propaganda; it falls so trippingly off so many different tongues because it works, igniting a host of historically conditioned fears all at the same time: fears of the Nazis, fears of being sterilized, fears of genocide, fears of being discriminated against, fears of being forced to do repugnant things, fears of being treated like a pawn in someone else's game of chess.

When the history of modern genetic screening is contrasted with the history of eugenic practices, however, the rationale for all those criticisms crumbles and such fears can begin to dissipate. Medical genetics was not created by eugenicists. The technological systems through which genetic screening is done were not developed to achieve eugenic goals. With the exception of some African-American communities in the 1970s, every community that has been affected by genetic screening has accepted it—even when screening has been mandated by a government. Except possibly in China, no current government requires women to undergo termination of a pregnancy after prenatal diagnosis—and even the Chinese exception is not due to direct mandate, but is an indirect consequence of the one-child policy. Everywhere else, invasive diagnostic tests are elective and so are the consequent abortions; even where the tests have become routine aspects of prenatal care, women can and do refuse either to allow them or to follow up with abortion. Discrimination against the disabled in schooling, housing, and employment has diminished, even though genetic screening for disabling conditions has become commonplace.

Furthermore, when the development of testing systems is examined, the complex moral system on which they were built becomes evident—and the claim that genetic screening is a fundamentally evil enterprise evaporates. Genetic screening responds to and supports the supremely ethical goal of reducing individual suffering. It also responds to and supports the moral code that inspires parents to protect their children from harm and from suffering.

Although the developers of these systems may not have had these goals in mind, consumers of genetic services certainly understood that genetic screening responds to and supports the moral dictates of feminism. For two centuries feminists have fought to increase women's choices in the moral expectation that free choice means freedom from oppression. The fight for suffrage was based on the notion that political liberty rests on the ability to make political choices, and the fight for married women's property rights was based on the assumption that economic liberty means free management of one's own financial resources. Those who fought for female education believed that full personhood could be achieved only through education, while those who fought for equal employment had the not-unrelated idea that women should have the same range of career choices as similarly situated men. And the fight for legal access to birth control and abortion was a fight to allow women to choose when, with whom, and under what conditions to become pregnant or to carry a pregnancy to term.

Eugenic practices are judged immoral today precisely because they interfered with that last set of choices, because they constrained reproductive choice by allowing one group of people to prohibit another group of people from reproducing. This is also, ironically, what the opponents of genetic screening would like to do. Opponents of abortion on any grounds would not allow women to choose when, with whom, and under what circumstances to bear a child. Disability rights activists and reproductive feminists would not allow parents to choose under what conditions, and with which fetus, to continue a pregnancy. Left-wing intellectuals would prefer that parents not make biologically based choices about the kind of children they wish to raise.

Genetic screening increases choice, and choice, as so many feminists have argued for so many years, is the sine qua non of freedom. Opponents of women's suffrage argued that political choices were too difficult for women's minds to make. Opponents of equal employment opportunity argued that it would create too many stresses for women who were also wives and mothers. Opponents of legal abortion say that women who choose to terminate pregnancies do not understand their own best interests. Free choice is often difficult, as many patients have discovered when a prenatal diagnosis is positive, but in this realm, as in the others, the difficulty of the choice is not a reason, in a free society, to prohibit people from making it. Proponents of genetic screening are, for all of these reasons, responding to and supporting the moral dictates of modern feminism.

Medical geneticists and genetic counselors are pursuing an ethical vision in their careers, opening up new, moral opportunities for their patients. They have no need to apologize for what was once done in the name of eugenics, because they are not now and never have been eugenicists, and because the opportunities they create for their patients are the very opposite of what eugenicists would have wanted. Women and men who seek out the services of medical geneticists and genetic counselors are also pursuing an ethical vision: of how and why they want to form their families. They should not be made to feel guilty about the choices they have freely made in their own interests and in the interests of their real and potential children.

Genetic screening increases reproductive choice, and it also provides hope, hope that many parents never had before—hope of having, not a perfect child, but a child who, at least at the start of life, is free of devastating disease or overwhelming disability. Screening has become a routine practice in many different countries, and in many different social circumstances, for precisely these good reasons. Negative consequences, for individuals or for the ethnic communities to which they belong or for the national societies of which they are citizens, have been rare. Surely, then, the time has come for prospective parents to stop feeling guilty about participating in screening, and for historians, social scientists, and journalists to stop warning about its hidden eugenic evils. Those evils do not exist—and continuing to insist that they do is an attempt to further a political agenda by making good people feel unnecessarily guilty about their fundamentally wise and moral behavior.

Notes

1. Many Varieties of Beautiful Inheritance

1. Francis Galton, "Hereditary talent and character," *Macmillan's Magazine* 12 (June 1865): 165.
2. Francis Galton, *Inquiries into human faculty* (London: Macmillan, 1883), 17.
3. Galton, "Hereditary talent and character," 165.
4. Georgina [Lady Theodore] Chambers, "Notes on the early days of the Eugenics Education Society" [1950], mss, Eugenics Society Papers, Wellcome Library, London, as quoted in Pauline M. H. Mazumdar, *Eugenics, human genetics and human failings: The Eugenics Society, its sources and its critics in Britain* (London: Routledge, 1992), 7.
5. From the text of the first involuntary sterilization statute, Indiana 1907, which was used as a model for all the others, as quoted in Philip Reilly, *The surgical solution: A history of involuntary sterilization in the United States* (Baltimore: Johns Hopkins University Press, 1991), 46.
6. Buck v. Bell, 274 U.S. 200 (1927), as quoted in Reilly, *The surgical solution,* 87.
7. Adolphe Landry, "L'eugénique," *Revue Bleu* 51 (1913): 782, as quoted in William H. Schneider, "The eugenics movement in France, 1890–1940," in Mark B. Adams, ed., *The Wellborn science: Eugenics in Germany, France, Brazil, and Russia* (New York: Oxford University Press, 1990), 75.
8. Aleksandr Serebrovskii, "Anthropogenetics and eugenics in a socialist

society," *Trudy kabineta* (1929): 18, as quoted in Mark Adams, "Eugenics in Russia," in Adams, *The Wellborn science,* 180–181.

9. Alfred Ploetz, *Die Tüchtigkeit unserer Rasse und der Schutz der Schwachen* (1895), as quoted in Robert Proctor, *Racial hygiene: Medicine under the Nazis* (Cambridge, MA: Harvard University Press, 1988), 21.
10. *Zeitschrift für Volksaufartung und Erbkunde* 1 (1926): 2, as quoted in Proctor, *Racial hygiene,* 47.
11. Theobald Lang, "Der Nationalsozialismus als politischer Ausdruck unserer biologischen Kenntnis," *Nationalsozialistische Monatshefte* 1 (1930): 393, as quoted in Proctor, *Racial hygiene,* 30.
12. Hermann Boehm, "Das erbbiologische Forschungsinstitut in Alt-Rehse," *Ärzteblatt für Berlin* 42 (1937): 415–416, as quoted in Proctor, *Racial hygiene,* 84.
13. Quoted in Proctor, *Racial hygiene,* 62; source unclear.
14. This was a secret directive; Proctor quotes this version from archival sources in *Racial hygiene,* 186.
15. Ibid., 132.
16. From Alfred Dorner, *Mathematik im Dienste der nationalpolitischen Erziehung* (1935), as quoted ibid., 184.
17. Proposed text of a sterilization law, from the Norwegian Consultative Eugenics Committee (1931), as quoted in Nils Roll-Hansen, "Norwegian eugenics: Sterilization as social reform," in Gunnar Broberg and Nils Roll-Hansen, *Eugenics and the welfare state: Sterilization policy in Denmark, Sweden, Norway, and Finland* (Lansing: Michigan State University Press, 1996), 172.
18. The text of the Convention can be found on the Council of Europe website, *conventions.coe.int/treaty/en/treaties/html/164.htm* (accessed 1 May 2007).
19. Penrose, memorandum, 4 Dec. 1961, as quoted in Daniel Kevles, *In the name of eugenics: Genetics and the uses of human heredity* (New York: Knopf, 1985), 252.

2. Eugenics and the Genealogical Fallacy

1. The letters remain private and unpublished; Alexander G. Bearn reports their existence and quotes from them in his biography of Garrod, *Archibald Garrod and the individuality of man* (Oxford: Oxford Uni-

versity Press, 1993), 59–62. The three publications are A. E. Garrod, "The incidence of alkaptonuria: A study in chemical individuality," *Lancet* (13 Dec. 1902): 1616–20; William Bateson and E. R. Saunders, *Report to the Evolution Committee of the Royal Society, I* (London: Harrison and Sons, 1902); and William Bateson, *Mendel's principles of heredity: A defence* (Cambridge: Cambridge University Press, 1902).

2. For a bibliography of Garrod's work, see Archibald E. Garrod, *Inborn errors of metabolism,* rpt. with additions by H. Harris (London: Oxford University Press, 1963), 97–109.
3. A. E. Garrod, "Alkaptonuria: A simple method for the extraction of homogentisic acid from the urine," *Journal of Physiology* 23 (1898–99): 512–514.
4. Garrod, "The incidence of alkaptonuria," 1617.
5. A. E. Garrod, "About alkaptonuria," *Lancet* (30 Nov. 1901): 1485.
6. Ibid, 1486.
7. Beatrice Bateson, *William Bateson, F.R.S., naturalist: His essays and addresses together with a short account of his life* (Cambridge: Cambridge University Press, 1928), 28, 62.
8. Mendel's paper was originally published in German: "Versuche über Planzen-Hybriden," *Verhandlungen des naturforschenden Vereines in Brünn* 4 (1865): 3–47. It was little read and never appreciated until three post-Darwinian botanists rediscovered it in 1900. Bateson read the paper in 1900, when it was republished by the Dutch naturalist Hugo de Vries, and immediately set about translating it into English as *Mendel's principles of heredity: A defence.*
9. Beatrice Bateson, *William Bateson,* 74.
10. William Bateson, "Common-sense in racial problems: The Galton Lecture," *Eugenics Review* (1919), rpt. in Beatrice Bateson, *William Bateson,* 371.
11. Garrod, "The incidence of alkaptonuria," 1619.
12. Bearn, *Archibald Garrod,* 99.
13. Ibid., 131, 132 (italics added).
14. Malcolm Kottler, "From 48 to 46: Cytological technique, preconception, and the counting of human chromosomes," *Bulletin of the History of Medicine* 48 (1974): 465–502.
15. Walter B. Sutton, "The chromosomes in heredity," *Biological Bulletin* 4 (1903): 241.
16. Sex had been one of the first traits studied by correlating microscopical

examination of chromosomes with the anatomical features of organisms. In 1905 Sutton's mentor at Columbia University, E. B. Wilson, and Nettie Stevens, a faculty member at Bryn Mawr College, had independently discovered that sex is determined by chromosomes; in mammals, they found, females have two copies of the X chromosome and males have one copy each of the X and the Y.

17. R. A. Fisher, "On the correlation between relatives on the supposition of Mendelian inheritance," *Transcript of the Royal Society of Edinburgh* 52 (1918): 399–433.
18. Bearn, *Archibald Garrod,* 145.
19. Karl Pearson, *Tuberculosis, heredity and environment: Eugenics Laboratory Lecture Series* (London: Galton Laboratory of National Eugenics, 1912), 45. Karl Pearson, *Eugenics and public health: Questions of the day and fray, VI* (London: Department of Applied Statistics, University College London, 1912), 32–33. Charles B. Davenport, "Heredity in medicine," in Charles B. Davenport et al., eds., *Medical genetics and eugenics* (Philadelphia: Women's Medical College of Philadelphia, 1940), 84.
20. Lewellys Barker, "Heredity in the clinic," *American Journal of the Medical Sciences* 173 (1927): 598, as quoted in Kenneth M. Ludmerer, *Genetics and American society: A historical appraisal* (Baltimore: Johns Hopkins University Press, 1972), 64.
21. J. B. S. Haldane, *Heredity and politics* (New York: Norton, 1937), 106, as quoted in Ludmerer, *Genetics and American society,* 66. Laurence Snyder, from an interview, as quoted in Daniel J. Kevles, *In the name of eugenics: Genetics and the uses of human heredity* (New York: Knopf, 1985), 210.
22. This and all the preceding quotations from Neel are from James V. Neel, *Physician to the gene pool: Genetic lessons and other stories* (New York: Wiley, 1994), 13, 17.

3. Pronatal Motives and Prenatal Diagnosis

1. See, e.g., F. Schatz, "Eine besondere Art von einseitiger Polyhydramnie mit anderseitiger Oligohydramnie bei eineiigen Zwillingen," *Archiv für Gynäkologie* 19 (1882): 329–369.
2. Various uses of amniotic taps are described in Seymour Romney et al., *Gynecology and obstetrics: The health care of women* (New York:

McGraw-Hill, 1975), 749–750. On the use of saline as an abortifacient, see Tanfer Emin-Tunc, "Technologies of choice: A history of abortion techniques in the United States, 1850–1980" (PhD diss., State University of New York at Stony Brook, 2005), ch. 4.

3. My accounts of the history of Rh disease diagnosis and therapy are drawn from James A. Stockman, "Overview of the state of the art of Rh disease: History, current clinical management, and recent progress," *Journal of Pediatric Hematology and Oncology* 23, no. 8 (Nov. 2001): 554–562, and David Zimmerman, *Rh: The intimate history of a disease and its conquest* (New York: Macmillan, 1975).
4. D. C. A. Bevis, "Preliminary communication: Composition of liquor amnii in haemolytic disease of the newborn," *Lancet* (30 Sept. 1950): 443; Bevis, "The antenatal prediction of haemolytic disease of the newborn," *Lancet* (23 Feb. 1952): 395–398; Bevis, "Composition of liquor amnii in haemolytic disease of the newborn," *Journal of Obstetrics and Gynaecology for the British Empire* 60 (1953): 244–251; Bevis, "Blood pigments in haemolytic disease of the newborn," ibid. 63 (1956): 68–75.
5. A. W. Liley, "Intrauterine transfusion of foetus in haemolytic disease," *British Medical Journal* (2 Nov. 1963): 1107–09.
6. H. Parrish, M. Rountree, and F. Lock, "Technique and experience with transabdominal amniocentesis in 50 normal patients," *American Journal of Obstetrics and Gynecology* (1958): 724–727; W. Cary, "Amniocentesis in haemolytic disease of the new-born," *Medical Journal of Australia* (1960): 778–781; A. W. Liley, "The technique and complications of amniocentesis," *New Zealand Journal of Medicine* 59 (1960): 581–586.
7. All the material about Hsu comes from his remarkably candid memoir, T. C. Hsu, *Human and mammalian cytogenetics: An historical perspective* (New York: Springer-Verlag, 1979).
8. On the history of the human chromosome number, see Hsu, *Human and mammalian cytogenetics,* ch. 2, as well as Malcolm Kottler, "From 48 to 46: Cytological technique, preconception and the counting of the human chromosomes," *Bulletin of the History of Medicine* 48 (1974): 465–502.
9. Hsu, *Human and mammalian cytogenetics,* 17 (italics added).
10. Ibid., 18.
11. T. C. Hsu, "Mammalian chromosomes *in vitro* I: The karyotype of

man," *Journal of Heredity* 43 (1952): 172; T. C. Hsu and C. M. Pomerat, "Mammalian chromosomes *in vitro* II: A method for spreading the chromosomes of cells in tissue culture," *Journal of Heredity* 44 (1953): 23–29.

12. J. H. Tijo and A. Levan, "The chromosome number of man," *Hereditas* 42 (1956): 1–6. C. E. Ford and J. L. Hamerton, "The chromosomes of man," *Nature* 178 (1956): 1020–23.
13. Three papers resulted from this research: R. Turpin and J. Lejeune, "Étude dermatoglyphique des paumes des mongoliens et de leurs parents et germaines," *La semaine des hôpitaux de Paris* 29 (14 Dec. 1953): 3955–67; R. Turpin and J. Lejeune, "Étude d'une famille comportant quatre frères et soeurs mongoliens," ibid., 3979–84; R. Turpin and J. Lejeune, "Analogies entre le type dermatoglyphique palmaire des singes inferieurs et celui des enfants atteints de mongolisme," *Les comptes rendus de l'Academie des sciences* 238 (18 Jan. 1954): 395–397. "Stigmates" is Lejeune's word.
14. Jérôme Lejeune, Marthe Gautier, and Raymond Turpin, "Étude des chromosomes somatiques de neuf enfants mongoliens," *Les comptes rendus de l'Academie des sciences* 248 (16 Mar. 1959): 1721. The translation, French to English, is mine. The ellipsis in this quotation is for the word *récente* and a footnote, which refers the reader to a previous article: ibid. 248 (26 Jan. 1959): 602–603, in which Lejeune and Gautier describe the culturing technique they used; Turpin's name appears on all articles published by members of his clinical staff.
15. Murray Barr, "Human cytogenetics—some reminiscences," *Bioessays* 9 (1988): 80.
16. Although Barr did not realize this for many years, the nuclear inclusion he was looking at was the inactivated X chromosome in the somatic cells of females; it stains clearly when the cell is at resting stage.
17. Murray Barr and E. G. Bertram, "A morphological distinction between neurons of the male and female, and the behavior of the nucleolar satellite during accelerated nucleoprotein synthesis," *Nature* 163 (1949): 676.
18. See M. L. Barr and G. E. Hobbs, "Chromosomal sex in transvestites," *Lancet* (29 May 1954): 1109–10; for citations to Barr's work on other chromosomal abnormalities of sex, see the endnotes to Barr, "Human cytogenetics."
19. K. L. Moore and M. L. Barr, "Smears from oral mucosa in detection of chromosomal sex," *Lancet* (9 July 1955): 57–58.

20. Fritz Fuchs and Povl Riis, "Antenatal sex determination," *Nature* 177 (18 Feb. 1956): 330. D. M. Serr, L. Sachs, and M. Danon, "Diagnosis of sex before birth using cells from the amniotic fluid," *Bulletin of the Research Council of Israel* 5B (Dec. 1955): 137; L. Sachs, D. M. Serr, and M. Danon, "Analysis of amniotic fluid cells for diagnosis of fetal sex," *British Medical Journal* 2 (1956): 795–798. E. L. Makowski, K. A. Prem, and I. H. Kaiser, "Detection of sex of fetuses by the incidence of sex chromatin body in nuclei of cells in amniotic fluid," *Science* 123 (30 Mar. 1956): 542–543; Landrum Shettles, "Nuclear morphology of cells in human amniotic fluid in relation to sex of infant," *American Journal of Obstetrics and Gynecology* 71, no. 4 (Apr. 1956): 834–838.
21. Fuchs and Riis, "Antenatal sex determination," 330. Sachs, Serr, and Danon, "Analysis of amniotic fluid," 798.
22. Author's interview with Fritz Fuchs, New York, 3 Mar. 1989.
23. Povl Riis and Fritz Fuchs, "Antenatal determination of foetal sex in prevention of hereditary diseases," *Lancet* (23 July 1960): 181.
24. Ibid.
25. Ibid, 180.
26. Ibid, 181 (italics added).
27. Fuchs interview.
28. Donald quoted, without citation, in Joseph Woo, "A short history of the development of ultrasound in obstetrics and gynecology," part 1, *www.ob-ultrasound.net/history1.html* (accessed Dec. 2006).
29. *Life* (10 Sept. 1965): 64.
30. Marshall D. Levine and Michael M. Kaback, "Prenatal genetic diagnosis and ultrasound—A perspective: 1975," typescript; Archives of National Institute of Child Health and Human Development (NICHD), NIH.
31. Fritz Fuchs, "Genetic information from amniotic fluid constituents," *Clinical Obstetrics and Gynecology* 9 (June 1966): 565–573.
32. Mark W. Steele and W. Roy Breg, "Chromosome analysis of human amniotic-fluid cells," *Lancet* (19 Feb. 1966): 385.
33. For reviews of this literature, see Fritz Fuchs and Lars J. Cederqvist, "Recent advances in antenatal diagnosis by amniotic fluid analysis," *Clinical Obstetrics and Gynecology* 13 (Mar. 1970): 178–201, and H. L. Nadler and A. B. Gerbie, "Role of amniocentesis in the intrauterine detection of genetic disorders," *New England Journal of Medicine* 282 (1970): 596–599.

34. D. J. H. Brock, *Early diagnosis of fetal defects* (London: Longman, 1982), ch. 4.
35. The earlier figure comes from "Draft proposal for amniocentesis registry" (21 Mar. 1971), typescript, Archives of NICHD, NIH. The later figure is from NICHD National Registry for Amniocentesis Study Group, "Midtrimester amniocentesis for prenatal diagnosis: Safety and accuracy," *JAMA* 236 (27 Sept. 1976): 1471.
36. NICHD National Registry for Amniocentesis Study Group, "Midtrimester amniocentesis for prenatal diagnosis," 1471. See also Barbara J. Culliton, "Amniocentesis: HEW backs test for prenatal diagnosis of disease," *Science* 190 (5 Nov. 1975): 537–540.
37. The first of these cases was Becker v. Schwartz (1978) 46NY 2nd 401, 386NE 2nd 807. See also B. L. Bernier, "Mothers as plaintiffs in prenatal tort liability cases: Recovery for physical and emotional damages," *Harvard Women's Law Journal* 4, no. 1 (spring 1981): 43–71.
38. Brock, *Early diagnosis of fetal defects,* ch. 4.
39. See Woo, "Short history of the development of ultrasound in obstetrics and gynecology," parts 2–3.
40. For a summary, see Valerie M. Hudson and Andrea M. den Boer, *Bare branches: Security implications of Asia's surplus male population* (Cambridge, MA: MIT Press, 1984).
41. For a review of this literature, see Ruth Schwartz Cowan, "Aspects of the history of prenatal diagnosis," *Fetal Diagnosis and Therapy* 8, supp. 1 (Feb. 1993): 10–17.
42. For an example of the use of prenatal diagnosis in arguing for abortion reform, see Sager C. Jain and Laurel Gooch, *Georgia Abortion Act, 1967: A study in legislative process* (Chapel Hill: University of North Carolina Population Center, 1969), 63.
43. Kristen Luker, *Abortion and the politics of motherhood* (Berkeley: University of California Press, 1984), esp. ch. 3.
44. Jérôme Lejeune, "The William Allan White Memorial Lecture: On the nature of men," *Annals of Human Genetics* 22 (Mar. 1970): 121–128. "Senate begins hearings on bill to outlaw abortions," *New York Times* (24 Apr. 1981): A16; "Commentary: The rights of infants with Down's syndrome," *JAMA* 251, no. 2 (13 Jan. 1984): 229.
45. Carlo Valentine, Edward J. Shutter, and Thiele Keats, "Prenatal diagnosis of Down's syndrome," *Lancet* (27 July 1968): 220 (italics added).
46. R. R. Gordon (letter), "Genetic therapeutic abortion," *Lancet* (5 Mar. 1966): 544.

47. U. Lange beck (letter), "Genetic therapeutic abortion," *Lancet* (14 May 1966): 1103. Bill Muehlenberg, review of Melinda Tankard Reist, *Defiant birth: Women who resist medical eugenics,* at *www.bill muehlenberg.com* (accessed Oct. 2006).

4. No Matter What, This Has to Stop!

1. A. Følling, "The original detection of phenylketonuria," in H. Bickel, F. P. Hudson, and L. I. Woolf, eds., *Phenylketonuria and some other inborn errors of amino acid metabolism* (Stuttgart: Georg Theme Verlag, 1971), 1–3.
2. A. Følling, "Uber Ausscheidung von Phenylbrenztraubensaure in den Harn als Stoffwechselanomalie in Verbindung mit Imbezilität," *Zeitschrift für physiologische Chemie* 227 (1934): 169–176.
3. Daniel J. Kevles, *In the name of eugenics: Genetics and the uses of human heredity* (New York: Knopf, 1985), 155.
4. L. S. Penrose, "Two cases of phenylpyruvic amentia," *Lancet* (5 Jan. 1935): 23–24; L. S. Penrose, "Inheritance of phenylpyruvic amentia (phenylketonuria)," *Lancet* (27 July 1935): 192–194.
5. "Memoirs—1964: Phenylketonuria," Lionel S. Penrose Papers, University College London, as quoted in Kevles, *In the name of eugenics,* 159.
6. H. Bickel, "The first treatment of phenylketonuria," *European Journal of Pediatrics* 155, supp. 1 (1996): S2; also H. Bickel, J. Gerrard, and E. Hickman, "Influence of phenylalanine intake on phenylketonuria," *Lancet* (17 Oct. 1953): 812–813.
7. This account is based on Samuel Bessman and Judith P. Swazey, "Phenylketonuria: A study of biomedical legislation," in E. Mendelsohn, Judith P. Swazey, and Irene Tavis, eds., *Human aspects of biomedical innovation* (Cambridge, MA: Harvard University Press, 1971), 49–76, and a memo to the authors from the Children's Bureau quoted in the notes, 55–56.
8. R. Guthrie, "The introduction of newborn screening for phenylketonuria," *European Journal of Pediatrics* 155, supp. 1 (1996): S4–S5.
9. This account of the history of NARC—including the quotations—is taken from Robert Segal, "The National Association for Retarded Citizens," at *www.thearc.org/history/segal.htm* (accessed 17 Dec. 2003).
10. David Goode, "The History of AHRC," *www.ahrcnyc.org/about/history.htm* (accessed 18 Dec. 2003), 44–45. AHRC (Association for

the Help of Retarded Children), founded at about the same time as NARC, is focused on the needs of retarded children and adults in New York City.

11. Robert Guthrie, "Newborn screening: Past, present and future," speech presented at the Birth Defects Symposium, Albany, NY, 1 Oct. 1985, copy in the records of the U.S. Children's Bureau, cited in Diane B. Paul, "The history of newborn phenylketonuria screening in the U.S.," in Neil A. Holtzman and Michael S. Watson, eds., *Promoting safe and effective genetic testing in the United States: Final report of the Task Force on Genetic Testing* (Baltimore: Johns Hopkins University Press, 1998), 139.
12. R. Guthrie, "Organization of a regional newborn screening laboratory," in H. Bickel, R. Guthrie, and G. Hammersen, eds., *Neonatal screening for inborn errors of metabolism* (Berlin: Springer-Verlag, 1980), 264, fig. 2.
13. Guthrie, "The introduction of newborn screening," S5.
14. Ibid.
15. R. Guthrie and A. Susi, "A simple phenylalanine method for detecting phenylketonuria in large populations of newborn infants," *Pediatrics* 32 (1963): 338–343.
16. On these and other initial objections, see Bessman and Swazey, "Phenylketonuria." The early initiation of dietary therapy was not conclusively shown to be helpful (in moderating the deficits, not in eliminating them completely) until the results of the United States Collaborative Study of Children Treated for Phenylketonuria were published in 1981: M. L. Williamson, R. Koch, C. Azen, and C. Chang, "Correlates of intelligence test results in treated phenylketonuric children," *Pediatrics* 68 (1981): 161–167. The fundamental causal mechanism that connects metabolic deficiency with neurological damage is still unknown.
17. On the issue of informed consent for newborn screening, see Diane Paul, "Contesting consent: The challenge to compulsory neonatal screening for PKU," *Perspectives in Biology and Medicine* 42, no. 2 (1999): 207–218.
18. In the late 1990s, however, the Supreme Court of Ireland ruled that parents have the right to refuse to have their child's heel pricked; that the authority of the parents overrides that of public health officials. The parents in this case objected to the use of blood, not to the diagnostic test. See G. Laurie, "Better to hesitate at the threshold of com-

pulsion: PKU testing and the concept of family autonomy in Eire," *Journal of Medical Ethics* 28 (2002): 136–138.

19. Warren Tay, "Symmetrical changes in the region of the yellow spot in each eye on an infant," *Transactions of the Ophthalmological Society of the United Kingdom* 1 (1881): 55–57. Warren Tay, "Third instance in same family of symmetrical changes in the region of yellow spot in each eye of an infant closely resembling those of embolism," ibid. 4 (1884): 158.
20. Bernard Sachs, "On arrested cerebral development, with special reference to its cortical pathology," *Journal of Nervous and Mental Disease* 14 (1887): 541–552.
21. Ibid., 550.
22. Bernard Sachs, "A family form of idiocy, generally fatal, associated with early blindness: Amauratic family idiocy," *Journal of Nervous and Mental Disease* 21 (1896): 475–479.
23. For a summary of these views, see David Slome, "The genetic basis of amaurotic family idiocy," *Journal of Genetics* 27 (Aug. 1933): 363–372.
24. Ibid, 369.
25. E. Klenk, "Beiträge zur Chemie der Lipoidosen: Niemann-Picksche Krankheit und amaurotische Idiotie," *Zeitschrift für physiologische Chemie* 262 (1939–40): 128–143; E. Klenk, "Über die Ganglioside des Gehirns bei der infantile amaurotischen Idiotie vom Typ Tay-Sachs," *Berichte der deutschen chemischen Gesellschaft* 75 (1942): 1632–36.
26. Author's interview with Ruth Dunkell, founding president of NTSAD, New York, June 2003. NTSAD is now known as the National Tay-Sachs and Allied Diseases Association.
27. See, e.g., Bruno Volk and Samuel Aronson, eds., *Sphingolipids, sphingolipidoses and allied disorders* (New York: Plenum, 1972); *Program of the gala benefit dinner dance to mark the 45th anniversary of NTSAD, 27 April 2002* (unpaginated); Ori Z. Soltes, *Reach for the moon: Ruth Dunkell* (exhibit catalog); both in the author's possession.
28. Lars Svennerholm, "The chemical structure of normal human brain and Tay-Sachs gangliosides," *Biochemical and Biophysical Research Communications* 9 (1962): 436–441.
29. Shintaro Okada and John S. O'Brien, "Tay-Sachs disease: Generalized absence of a beta-D-N-acetylhexosaminidase component," *Science* 165 (1969): 698–700.

30. Author's interview with Sam Dunkell, founding chair of the Medical Advisory Board of NTSAD, New York, June 2003.
31. L. Schneck, C. Valenti, et al., "Prenatal diagnosis of Tay-Sachs disease," *Lancet* (21 Mar. 1970): 582–584. J. S. O'Brien, S. Okada, et al., "Tay-Sachs disease: Prenatal diagnosis," *Science* 172 (2 Apr. 1971): 61–64.
32. Ruth Dunkell interview, June 2003.
33. All quotations from Michael M. Kaback, "Screening and prevention in Tay-Sachs disease: Origins, update and impact," in Robert J. Desnick and Michael M. Kaback, eds., *Tay-Sachs disease* (San Diego: Academic Press, 2001), 254–255.
34. See, e.g., M. M. Kaback and R. S. Zeiger, "Heterozygote detection in Tay-Sachs disease: A prototype community genetics screening program for the prevention of recessive genetic disorders," in Volk and Aronson, *Sphingolipids,* 613–632; and M. M. Kaback et al., "Approaches to the control and prevention of Tay-Sachs disease," *Progress in Medical Genetics* 10 (1974): 103–134.
35. Kaback, "Screening and prevention," 259.
36. This controversy briefly spilled over into the medical press; see the exchange of letters that followed the publication of M. D. Kuhr, "Doubtful benefits of Tay-Sachs screening," *New England Journal of Medicine* 292 (1975): 371.
37. Josef Ekstein and Howard Katzenstein, "The Dor Yeshorim story: community-based carrier screening for Tay-Sachs disease," in Desnick and Kaback, *Tay-Sachs disease,* 299. This quotation does not refer to the much larger group of Modern Orthodox Jews, many of whom were willing to consider prenatal diagnosis and abortion in the special case of Tay-Sachs families. For more on this distinction, see Fred Rosner, "Tay-Sachs disease: To screen or not to screen," *Journal of Religion and Health* 15 (1976): 271–280.
38. Alison George, "The rabbi's dilemma," *www.newscientist.com/opinion/opinioninterview.jsp?id=ns24341* (accessed 14 June 2004), 2.
39. Ekstein and Katzenstein, "The Dor Yeshorim story," 307.
40. For a summary, see Katherine L. Acuff and Ruth R. Faden, "A history of prenatal and newborn screening programs: Lessons for the future," in Ruth R. Faden et al., *AIDS, women and the next generation: Towards a morally acceptable public policy for HIV testing of pregnant women and newborns* (New York: Oxford, 1991), 59–93.
41. Kaback, "Screening and prevention," 260.

5. Genetic Screening and Genocidal Claims

1. James Herrick, "Peculiar elongated and sickle-shaped red blood cells in a case of severe anemia," *Transactions of the Association of American Physicians* 25 (1910): 553–561, and "Peculiar elongated and sickle-shaped red blood corpuscles in a case of severe anemia," *Archives of Internal Medicine* 6 (1910): 517–521.
2. Victor Emmel, "A study of the erythrocytes in a case of severe anemia with elongated sickle-shaped red blood corpuscles," *Archives of Internal Medicine* 20 (1917): 592.
3. Verne Mason, "Sickle cell anemia," *JAMA* 79 (1922): 1318–19; John Huck, "Sickle cell anemia," *Bulletin of the Johns Hopkins Hospital* 34 (1923): 335–344; W. Taliaferro and J. Huck, "Inheritance of sickle cell anemia in man," *Genetics* 8 (1923): 594–598.
4. M. A. Ogden, "Sickle cell anemia in the white race," *JAMA* 71 (1943): 164–182, as cited in Keith Wailoo, *Drawing blood: Technology and disease identity in twentieth-century America* (Baltimore: Johns Hopkins University Press, 1997): 137.
5. For a dissenting view, see Thomas Cooley, "Hereditary factors in the blood dyscrasias," *American Journal of Diseases of Children* 62 (1941): 1–7.
6. J. V. Neel and W. N. Valentine, "Hematologic and genetic study of the transmission of thalassemia," *Archives of Internal Medicine* 74 (1944): 185–196.
7. James Neel, *Physician to the gene pool: Genetic lessons and other stories* (New York: Wiley, 1994), 43.
8. James Neel, "The inheritance of sickle cell anemia," *Science* 110 (1949): 64–66.
9. Neel, *Physician to the gene pool,* 43. This same discovery was made independently by a researcher in Africa; see E. A. Beet, "The genetics of the sickle cell trait in a Bantu tribe," *Annals of Eugenics* 14 (1949): 279–284.
10. Linus Pauling et al., "Sickle cell anemia: A molecular disease," *Science* 110 (1949): 543–548.
11. Much of my understanding of clinical approaches to sickle-cell disease between 1950 and 1970 comes from Keith Wailoo, *Dying in the city of the blues: Sickle cell anemia and the politics of race and health* (Chapel Hill: University of North Carolina Press, 2001).

12. Z. O. Webb, "Sickle cell anemia: A clinical screening survey," *Journal of the National Medical Association* 64, no. 3 (1972): 197–199.
13. D. M. Canning and R. G. Huntsman, "An assessment of Sickledex as an alternative to the sickling test," *Journal of Clinical Pathology* 23, no. 8 (Nov. 1970): 736–737.
14. Yvette F. Francis, Doris Wethers, and Lila Fenwick, "The Foundation for Research and Education in Sickle Cell Disease: A prospectus," *Journal of the National Medical Association* 62, no. 3 (1970): 200, 201. Francis was director of the sickle cell clinic at Jamaica Hospital, and Wethers was director of pediatrics at Sydenham and Knickerbocker Hospitals, all in New York City. Fenwick was the director of the foundation. Internal evidence in the article suggests that the foundation was formed in the mid-1960s.
15. Ibid., 201.
16. Robert B. Scott, "Health care priority and sickle cell anemia," *JAMA* 214 (26 Oct. 1970): 731–734. "Editorial: Sickle cell anemia," ibid., 749.
17. Nancy Hicks, "Doctor asks curb of Negro disease: Terms sickle cell anemia major health concern," *New York Times* (27 Oct. 1970): 51. Scott, "Health care priority," 734. The editors of *JAMA* note that Dr. Roland B. Scott (no relation to Robert B. Scott), who was chair of the department of pediatrics at Howard University, had been making the same argument for twenty years: see Roland B. Scott, "A commentary on sickle cell disease: The need for greater financial support and wider involvement of black professionals in a dynamic program to improve health services for victims of this malady," *Journal of the National Medical Association* 63 (Jan. 1971): 1–2, 60.
18. Barbara J. Culliton, "Sickle cell anemia: National program raises problems as well as hopes," *Science* 178 (20 Oct. 1972): 283–286.
19. Phillip J. Reilly, "Sickle cell anemia legislation, I," *Journal of Legal Medicine* (Sep.–Oct. 1973): 39–48; Reilly, "Sickle cell anemia, II," ibid. (Nov.–Dec. 1973): 36–40.
20. This account comes from Reilly, "Sickle cell anemia legislation," and from Lawrence E. Gary, "The sickle cell controversy," *Social Work* (May 1974): 263–272.
21. Pearson, as quoted in Culliton, "Sickle cell anemia: National program," 283.
22. P. T. Rowley, "Newborn screening for sickle-cell disease: benefits and burdens," *New York State Journal of Medicine* 78, no. 1 (Jan. 1978): 42–44.

23. D. B. Kellon and E. Buetler, "Physician attitudes about sickle cell disease and sickle cell trait" (editorial), *JAMA* 227 (1974): 71–72.
24. W. C. Mentzer Jr. et al., "Screening for sickle-cell trait and G-PD deficiency," *New England Journal of Medicine* 282, no. 20 (14 May 1970): 1155–56. Francis et al., "Foundation for Research and Education in Sickle Cell Disease," 203. For some of the pre-1970 literature on the hazards of sickle-cell trait, see W. A. Miller et al., "Perirenal hematoma in association with renal infarction in sickle-cell trait," *Radiology* 92 (1969): 351–352.
25. Stephen R. Jones et al., "Sudden death in sickle cell trait," *New England Journal of Medicine* 282, no. 6 (5 Feb. 1970): 323–325. To assess the medical controversy, see the Correspondence Section of *NEJM* from mid-February to late November 1970. Arthur H. Coleman and James R. Abernathy, "Sickle cell anemia: A legal conspectus," *Journal of the National Medical Association* 64, no. 6 (Nov. 1973): 354–355.
26. Barbara J. Culliton, "Sickle cell screening of recruits urged," *Science* 179 (16 Feb. 1973): 663.
27. Coleman and Abernathy, "Sickle cell anemia," 355. Richard Severo, "Air Academy to drop its ban on applicants with sickle cell gene," *New York Times* (4 Feb. 1981), accessed through *query.nytimes.com,* 3 July 2006.
28. Mary L. Hampton et al., "Sickle cell 'nondisease': A potentially serious public health problem," *American Journal of Diseases of Children* 128 (July 1974): 58–61.
29. As quoted in Barbara Culliton, "Sickle cell anemia: The route from obscurity to prominence," *Science* 178 (13 Oct. 1972): 142.
30. On the NIH, see Culliton, "Sickle cell anemia: National program," 284.
31. On black views from the 1920s to the 1970s, see Robert G. Weisbord, *Genocide? Birth control and the black American* (Westport, CT: Greenwood, 1975).
32. Carole R. McCann, *Birth control politics in the United States, 1916–1945* (Ithaca: Cornell University Press, 1994), esp. ch. 5.
33. Daniel H. Watts, "Birth control," *Liberator* (May 1969), 3, as quoted in Robert G. Weisbord, "Birth control and the black American: A matter of genocide?" *Demography* 10, no. 4 (Nov. 1973): 577. Congressional Black Caucus, U.S. House of Representatives, *A position on health in the black community* (Washington, 1972), 9. Dick Gregory, "My answer to genocide," *Ebony* 26, no. 12 (Nov. 1971): 66; see the December issue for the letters of comment.

34. *Ebony* 26, no. 12 (Dec. 1971): 16–17. Shirley Chisholm, *Unbought and unbossed* (Boston: Houghton Mifflin, 1970), 114. On the ambivalence and the powerlessness of women in the black power movements, see Bettye Collier-Thomas and V. P. Franklin, eds., *Sisters in the struggle: African-American women in the civil rights–black power movement* (New York: New York University Press, 2001), esp. the articles by Cynthia Griggs Fleming, Farah Jazsmine Griffin, and Tracy A. Matthews.
35. Alyce C. Gullattee, "Medico-legal insurance implications of sickle cell anemia," *Journal of the National Medical Association* 65, no. 3 (Sept. 1973): 416 (italics added).
36. Congressional Black Caucus, *A position,* 5.
37. "Illinois Chapter, Black Panther Party: Serving the people body and soul," *Black Panther* (10 Apr. 1971): 4.
38. *Black Panther* (1 May 1971): 12.
39. "The sickle cell game," ibid. (27 May 1972): 11.
40. *Black Panther* (10 Apr. 1971): 3; (26 Feb. 1972): 9.
41. Ibid. (19 Aug. 1972): 3, 9, 10.
42. Rayna Rapp, *Testing women, testing the fetus* (New York: Routledge, 2000), 165–166; L. A. Furr, "Perceptions of genetics research as harmful to society: Differences among samples of African-Americans and European-Americans," *Genetic Testing* 6, no. 1 (spring 2002): 563–572; S. L. Laskey et al., "Attitudes of African American premedical students toward genetic testing and screening," *Genetics in Medicine* 5, no. 1 (Jan./Feb. 2003): 49–54.

6. Parents, Politicians, Physicians, and Priests

1. E. Baysal et al., "The beta-thalassaemia mutations in the population of Cyprus, *British Journal of Haematology* 81, no. 4 (Aug. 1992): 607–609.
2. P. W. Dill-Russell, *Annual medical and sanitary report, 1955* (Nicosia: Cyprus Government Printing Office, 1956), 1.
3. Thomas B. Cooley and Pearl Lee, "A series of cases of splenomegaly in children with anemia and peculiar bone changes," *Transactions of the American Pediatric Society* 37 (1925): 29–30. Decades later, after the chemical structure of hemoglobin was revealed and the genes that contribute to making it were identified, hematologists came to understand that there are several thalassemias, what are now called the

"thalassemia syndromes." The one Cooley and Lee identified is now called homozygous β-thalassemia.

4. Thomas B. Cooley and Pearl Lee, "Erythroblastic anemia: Additional comments," *American Journal of Diseases of Children* 43 (1932): 706.
5. The story of the name is told in D. J. Weatherall and J. B. Clegg, *The β-thalassemia syndromes,* 3rd ed. (London: Blackwell, 1979), 3–5. The interchangeability of these names is illustrated on the website for the U.S.-based Cooley's Anemia Foundation, which supports patients, families, and research by "Leading the fight against β-thalassemia": *www.cooleysanemia.org.*
6. Shortly after her father's death, Cooley's daughter Emily recalled: "He became so annoyed by some of the ill-digested theories of racial limitation of characteristics . . . that he spent a good deal of time pointing out that . . . mutations can occur spontaneously in many populations." Emily H. Cooley to George White, 11 May 1948, Children's Hospital of Detroit Papers, Wayne State University Archives of Labor and Urban Affairs.
7. J. Caminopetros, "Recherches sur l'anémie erythroblastique infantile, des peoples de la Méditerranée Orientale, I," *Annales de médecine* 43 (Jan. 1938): 27–43; and "Recherches sur l'anémie erythroblastique infantile, des peoples de la Méditerranée Orientale, II," ibid. (Feb. 1938): 104–125.
8. For a review of these studies, see Weatherall and Clegg, β-thalassemia *syndromes,* 6. On the Italian literature, see Stefano Canali, "From splenic anemia in infancy to microcythemia: The Italian contribution to the description of the genetic bases of β-thalassemia," *Medicina nei secoli* 17, no. 1 (2005): 161–179. J. V. Neel and W. N. Valentine, "Hematologic and genetic study of the transmission of β-thalassemia," *Archives of Internal Medicine* 74 (1944): 185–196.
9. Allan Fawdry, "Report on the present state of knowledge of erythroblastic anaemia in Cyprus," in *Annual medical and sanitary report, 1946* (Nicosia), appendix D; rpt. in Bernadette Modell and Vasili Berdoukas, *The clinical approach to thalassemia* (London: Grune and Stratton, 1984), 267.
10. The summary that follows is based on my reading of the papers in the *First* (1963), *Second* (1969), *Third* (1973), *Fourth* (1980), and *Fifth* (1985) *Conferences on the Problems of Cooley's Anemia,* published as volumes of the *Annals of the New York Academy of Sciences.*

11. On the prospects for oral chelation, see J. C. Barton, "Drug evaluation: Deferitrin (GT-56–252; NaHBED) for iron overload disorders," *IDrug* 10, no. 4 (Apr. 2007): 270–281; Dominique Tobbell, "Charitable innovations: The political economy of thalassemia research and drug development in the United States, 1960–2000," in Judy Slinn and Vivianne Quirke, eds., *Perspectives on twentieth-century pharmaceuticals* (Bern: Peter Lang, 2007); and Miriam Shuchman, *The drug trial: Nancy Olivieri and the scandal that shook the Hospital for Sick Children* (Toronto: Random House, 2005).
12. Overheard by the author, Nicosia, April 1999.
13. Letter, Allan Fawdry to Michael Angastiniotis [1992], copy given to the author by Angastiniotis.
14. Allan Fawdry, "Report on the present state of knowledge of erythroblastic anaemia in Cyprus," *Transactions of the Royal Society of Tropical Medicine and Hygiene* 40 (1946): 87. Although this article bears the same title as the one cited above, the texts are considerably different.
15. Fawdry, "Report," in *Annual medical and sanitary report,* 12, and in Modell and Berdoukas, *Clinical approach,* 268.
16. Author's interview with Michael Angastiniotis, Nicosia, March 1999. Fawdry, "Report," in *Annual medical and sanitary report,* 15, and in Modell and Berdoukas, *Clinical approach,* 266–267.
17. T. Ashiotis, G. Stamatoyannopoulos, et al., "Thalassaemia in Cyprus," *British Medical Journal* 7, no. 2 (Apr. 1973): 38–42.
18. Author's interview with Minas Hadjiminas, Nicosia, March 1999.
19. Author's interviews with Ayten Berkalp, Lefkoşa, March 1999, and Colin Bate, Wigan, England, July 2001.
20. *Cyprus Antianaemic Society Bulletin,* no. 1 (1974) [unpaginated]; copy given to the author by Michael Angastiniotis.
21. Author's interviews with Hadjiminas, Angastiniotis, and Panos Englezos, Nicosia, spring 1999. Panos Englezos is a Nicosia businessman, parent of a son who had β-thalassemia, and long-term president of the Thalassemia International Foundation.
22. Author's interviews with Gülsen Bozkurt and Ayten Berkalp, Lefkoşa, March and April 1999.
23. G. Stamatoyannopoulos, "Report to Cyprus government: EM/Hum. Genet/2 Cyprus 8501/Rex 0028" [1972], typescript, copy given to the author by Michael Angastiniotis. Some of Stamatoyannopoulos's con-

clusions were published as part of Ashiotis and Stamatoyannopoulos et al., "Thalassaemia in Cyprus." According to Ayten Berkalp, Bernadette Modell visited Northern Cyprus in 1979 to give the same advice to its new ministry of health; Berkalp interview, March 1999.

24. Hadjiminas interview, March 1999.
25. M. Hadjiminas and M. Angastiniotis, "Prevention of thalassaemia in Cyprus," *Lancet* (14 Feb. 1981): 369–371.
26. Michael Angastiniotis, "Factors limiting the effective delivery of fetal diagnosis for thalassaemia" [n.d., probably 1981], typescript, copy given to the author by Angastiniotis.
27. Hadjiminas and Angastiniotis interviews, spring 1999. See also M. Angastiniotis, S. Kyriakidou, and M. Hadjiminas, "The Cyprus β-thalassemia control program," *Birth Defects, Original Article Series* 23, no. 5B (1988): 417–432, which is the source of the figures in this and the preceding paragraph.
28. Y. W. Kan, C. Valenti, et al., "Fetal blood sampling in utero," *Lancet* (19 Jan. 1974): 79–80; J. C. Hobbins and M. G. Mahoney, "In utero diagnosis of haemoglobinopathies: Technic for obtaining fetal blood," *New England Journal of Medicine* 290 (1974): 1065–68.
29. B. Modell, R. H. T. Ward, and D.V. I. Fairweather, "Effect of introducing antenatal diagnosis on reproductive behaviour of families at risk for thalassaemia major," *British Medical Journal* 280 (7 June 1980): 1347–50 (italics added).
30. Angastiniotis et al., "Cyprus β-thalassemia control program," 419.
31. Angastiniotis, "Factors limiting the effective delivery of fetal diagnosis for thalassaemia," p. 4 and table 5.
32. Hadjiminas interview, March 1999.
33. Angastiniotis, "Factors limiting," 5; Hadjiminas interviews, spring 1999.
34. "Northern Cyprus Thalassaemic Association & thalassaemia control program," typescript of a presentation made by Ayten Berkalp at the Second Mediterranean Meeting on Thalassaemia, Milan, Nov. 1985; copy given to the author by Berkalp.
35. Ibid., 3.
36. Gülsen Bozkurt, "Concerted action on developing patient registers as a tool for improving service delivery for haemoglobin disorders: Report from Northern Cyprus" [n.d., probably 1992], typescript, copy given to the author by Bozkurt.

37. Hadjiminas interview, April 1999.
38. Hadjiminas and Englezos interviews, spring 1999.
39. Hadjiminas and Angastiniotis, "Prevention of thalassaemia in Cyprus."
40. WHO Working Group, "Community control of hereditary anaemias: Memorandum from a WHO meeting," *Bulletin of the World Health Organization* 61 (1983): 63–80.
41. Angastiniotis et al., "The Cyprus β-thalassemia control program," 429.
42. R. Hoedemaekers and H. ten Have, "Geneticization: The Cyprus paradigm," *Journal of Medicine and Philosophy* 23, no. 3 (1998): 274–287.
43. Fawdry, "Report," in *Annual Medical and Sanitary Report,* 13, and in Modell and Berdoukas, *Clinical approach,* 67–68 (italics added).
44. Angastiniotis, "Factors limiting," 4.
45. Angastiniotis interview, April 1999.
46. *The Cyprus Antianaemic Society,* 4.
47. Author's interview with Voula Sarri, Nicosia, May 1999.
48. J. M. Old, R. H. Ward, et al., "First-trimester fetal diagnosis for haemoglobinopathies: Three cases," *Lancet* (25 Dec. 1982): 1413–16.
49. Author's interview with Maria Yiangou, Nicosia, May 1999. The same point was made by Ahmet Varoğlu, a thalassemia patient and chairman of the Northern Cyprus Thalassaemia Association, when interviewed by the author, Lefkoşa, April 1999. See also A. DiPalma et al., "Psychosocial integration of adolescents and young adults with thalassemia major," *Annals of the New York Academy of Sciences* 850 (30 June 1998): 355–360.
50. Medical entrepreneurs had wanted to open a PGD clinic on the island, largely to serve Saudi and Kuwaiti couples; the health ministry declined to license the clinic until it agreed to provide services free to Cypriot couples. Angastiniotis interview; author's interview with Anver M. Kuliev, Maroni, Cyprus, spring 1999.

Conclusion

1. The text of the Convention can be found on the Council of Europe website: *conventions.coe.int/treaty/en/treaties/html/164.htm* (accessed 1 May 2007).

2. Dorothy Nelkin and Laurence Tancredi, *Dangerous diagnostics: The social power of biological information* (New York: Basic Books, 1989), 161. Abby Lippman, "Worrying—and worrying about—the geneticization of reproduction and health," in Gwynne Basen et al., eds., *Misconceptions: The social construction of choice and the new reproductive and genetic technologies,* I (Hull, Quebec: Voyageur, 1993), 63.
3. Patricia Spallone, *Beyond conception: The new politics of reproduction* (Granby, MA: Bergin and Garvey, 1989), 113.
4. Barbara Katz Rothman, "Looking toward the future: Feminism and reproductive technologies: The James McCormick Mitchell Lecture," *Buffalo Law Review* 137 (1988/89): 214. Gena Corea, *The mother machine: Reproductive technologies from artificial insemination to artificial wombs* (New York: Harper and Row, 1985), 206.
5. Spallone, *Beyond conception,* 113.
6. See, e.g., Adrienne Asch, "Reproductive technology and disability," in Sherrill Cohen and Nadine Taub, eds., *Reproductive laws for the 1990s* (Clifton, NJ: Humana Press, 1989), 69–124; Adrienne Asch, "Why I haven't changed my mind about prenatal diagnosis," in Eric Parens and Adrienne Asch, eds., *Prenatal testing and disability rights* (Washington: Georgetown University Press, 2000), 234–258.
7. Marsha Saxton, "Disability rights and selective abortion," in Ricki Solinger, ed., *Abortion wars: A half century of struggle, 1950 to 2000* (Berkeley: University of California Press, 1998), 374–395.
8. Adrienne Asch, "Prenatal diagnosis and selective abortion: A challenge to practice and policy," *American Journal of Public Health* 89 (1999): 1655.
9. Bill Muehlenberg, review of Melinda Tankard Reist, *Defiant birth: Women who resist medical eugenics,* at *www.billmuehlenberg.com,* posted 17 May 2006 (accessed 22 May 2007).
10. Troy Duster, *Backdoor to eugenics* (New York: Routledge, 1990).
11. See Kenneth J. Ludmerer, *Genetics and American society* (Baltimore: Johns Hopkins University Press, 1972), ch. 3.
12. Y. W. Kan and A. M. Dozy, "Antenatal diagnosis of sickle cell anemia by DNA analysis of amniotic fluid cells," *Lancet* (1978): 910–912.
13. James V. Neel, *Physician to the gene pool: Genetic lessons and other stories* (New York: Wiley, 1994), 17.

14. Barbara Seaman, "How late can you wait to have a baby?" *Ms.* (Jan. 1976): 45–48, 78–79.
15. M. Neil MacIntyre, "Genetic risk, prenatal diagnosis and selective abortion," in David F. Walbert and Douglass Butler, eds., *Abortion, society and the law* (Cleveland: Case Western Reserve University Press, 1973), 237, 236, 227, 236.

Further Reading

Throughout this book I have tried to minimize the number and length of endnotes, providing just what seemed to be the minimum necessary for readers to check the evidence on which I have built my arguments. What follows is for readers who would like to know more about some of the subjects I have discussed, or who are curious about what other scholars, taking different approaches, have had to say on these subjects.

This is not by any means a complete bibliography of the historical, social, and political aspects of genetic screening. I have largely restricted myself to books in English that I have found helpful, insightful, and readable. The journal literature about genetic screening is vast, in medical periodicals as well as those focused on bioethics and on the social sciences, but because the electronic databases that now give readers access to journals are so easy to use, I have decided not to duplicate what they can do more comprehensively.

On the History of Eugenics

Most books on the history of eugenics focus on just one country, but there are some important exceptions: Diane B. Paul's two books, *Controlling Human Heredity, 1856 to the Present* (Humanities Press, 1995), and *The Politics of Heredity: Essays on Eugenics,*

Bio-medicine and the Nature-Nurture Debate (State University of New York Press, 1998); Daniel J. Kevles, *In the Name of Eugenics* (Knopf, 1985); Nancy Stepan, *The Hour of Eugenics: Race, Gender, and Nation in Latin America* (Cornell University Press, 1991); Mark B. Adams, *The Wellborn Science: Eugenics in Germany, France, Brazil and Russia* (Oxford University Press, 1990); Gunnar Broberg and Nils Roll-Hansen, eds., *Eugenics and the Welfare State: Sterilization Policy in Denmark, Sweden, Norway, and Finland* (Michigan State University Press, 1996).

American eugenics has received a great deal of historical attention. I have found several books particularly helpful and interesting: Harry Bruinius, *Better for All the World: The Secret History of Forced Sterilization and America's Quest for Racial Purity* (Knopf, 2006); Mark Haller, *Eugenics: Hereditarian Attitudes in American Society* (Rutgers University Press, 1963); Kenneth J. Ludmerer, *Genetics and American Society: An Historical Appraisal* (Johns Hopkins University Press, 1972); Steven Selden, *Inheriting Shame: The Story of Eugenics and Racism in America* (Teachers College Press, 1999); Philip Reilly, *The Surgical Solution: A History of Involuntary Sterilization in the United States* (Johns Hopkins University Press, 1991).

The most accessible books on British eugenics are Pauline M. H. Mazumdar, *Eugenics, Human Genetics and Human Failings: The Eugenics Society, Its Sources and Its Critics in Britain* (Routledge, 1992); G. R. Searle, *Eugenics and Politics in Britain, 1900–1914* (Nordhoff, 1976); and Matthew Thomson, *The Problem of Mental Deficiency: Eugenics, Democracy and Social Policy in Britain, 1870–1959* (Oxford University Press, 2002). The standard work in English on French eugenics is William H. Schneider, *Quality and Quantity: The Quest for Biological Regeneration in Twentieth Century France* (Cambridge University Press, 1990). On Soviet eugenics I have profitably read Mark B. Adams, "Eugenics in Russia, 1900–1940," in Adams, *The Wellborn Science;* and Loren R. Graham, "Science and Values: The Eugenics Movement in Germany and Russia in the 1920s," *American Historical Review* (1977).

On German eugenics I have found the following books most useful: Robert N. Proctor, *Racial Hygiene: Medicine under the Nazis* (Harvard University Press, 1988); Paul Weindling, *Health, Race and German Politics between National Unification and Nazism, 1870–1945* (Cambridge University Press, 1989); and Sheila Faith Weiss, *Race Hygiene and National Efficiency: The Eugenics of Wilhelm Schallmeyer* (University of California Press, 1987).

None of the biographical works on Galton is completely satisfactory, but Michael Bulmer, *Francis Galton: Pioneer of Heredity and Biometry* (Johns Hopkins University Press, 2003), and my own *Sir Francis Galton and the Study of Heredity in the Nineteenth Century* (Garland, 1985), are reasonable places to start. See also Elof Axel Carlson, *Genes, Radiation, and Society: The Life and Work of H. J. Muller* (Cornell University Press, 1981), and Ronald Clark, *JBS: The Life and Work of J. B. S. Haldane* (Coward McCann, 1968). For readers without mathematical facility, Joan Fisher Box's biography of her father, *R. A. Fisher: The Life of a Scientist* (Wiley, 1978), will be somewhat harder going. No full-scale biography of Karl Pearson or Charles Davenport has yet been written.

On the Many Facets of Genetics

The development of classical Mendelian genetics has received a fair amount of attention from historians of biology. The best introductions are in Jan Sapp, *Genesis: The Evolution of Biology* (Oxford University Press, 2003), part 3; Garland Allen, *Life Science in the Twentieth Century* (Cambridge University Press, 1978), ch. 4; and Peter Bowler, *The Mendelian Revolution: The Emergence of Hereditarian Concepts in Science and Society* (Johns Hopkins University Press, 1989). I also recommend Garland Allen, *Thomas Hunt Morgan: The Man and His Science* (Princeton University Press, 1978), and Robert E. Kohler, *Lords of the Fly: Drosophila Genetics and the Experimental Life* (University of Chicago Press, 1994).

Three classical geneticists have written histories of their field (and one of them has edited a collection of essays on the subject), which may be attractive to readers well versed in the genetic sciences: L. C. Dunn, *A Short History of Genetics: The Development of Some of the Main Lines of Thought, 1864–1939* (Iowa State University Press, 1991); L. C. Dunn, ed., *Genetics in the Twentieth Century: Essays on the Progress of Genetics during Its First Fifty Years* (Macmillan, 1951); Elof Axel Carlson, *The Gene: A Critical History* (Saunders, 1966); A. H. Sturtevant, *A History of Genetics* (Harper and Row, 1965).

The history of human genetics is wrapped up with the history of statistics and the history of evolutionary theory. Daniel J. Kevles attends to many aspects of the field in *In the Name of Eugenics,* as does Donald MacKenzie in *Statistics in Britain, 1865–1900: The Social Construction of Scientific Knowledge* (Edinburgh University Press, 1981). Some of the essays in Krishna R. Dronamraju, ed., *The History and Development of Human Genetics: Progress in Different Countries* (World Scientific, 1992) are insightful; the same is true for those in Milo Keynes, A. W. F. Edwards, and Robert Peel, eds., *A Century of Mendelism in Human Genetics* (CRC Press, 2004).

The history of medical genetics has been somewhat better served than that of human genetics. A good place to start is with James Neel's autobiographical memoir, *Physician to the Gene Pool: Genetic Lessons and Other Stories* (Wiley, 1994). I also recommend Susan Lindee, *Moments of Truth in Genetic Medicine* (Johns Hopkins University Press, 2005); Alan Rushton, *Genetics and Medicine in the United States, 1800 to 1922* (Johns Hopkins University Press, 1994); Diane Paul, *Controlling Human Heredity* and *The Politics of Heredity;* Diane Paul and Robert G. Resta, eds., "Historical Aspects of Medical Genetics," *American Journal of Medical Genetics, Part C: Seminars in Medical Genetics* 115 (2002). Several chapters in Kevles, *In the Name of Eugenics,* and in Daniel J. Kevles and LeRoy Hood, eds., *The Code of Codes: Scientific and Social Issues in the Human Genome Project* (Harvard University Press, 1992), also deal with the history of medical genetics. Sheldon Reed, one of the founders of genetic counseling, has written about his field in "A Short History of Genetic Counseling," *Social Biology* 21 (1974).

On Prenatal Diagnosis and Abortion

On the social meaning of various aspects of prenatal diagnosis and abortion, see Rayna Rapp, *Testing Women, Testing the Fetus: The Social Impact of Amniocentesis in America* (Routledge, 2000); Troy Duster, *Backdoor to Eugenics* (Routledge, 1990); Aliza Kolker and Meredith Burke, *Prenatal Testing: A Sociological Perspective* (Bergin and Garvey, 1994); Barbara Katz Rothman, *The Tentative Pregnancy: Prenatal Diagnosis and the Future of Motherhood* (Norton, 1986); Lisa M. Mitchell, *Baby's First Picture: Ultrasound and the Politics of Fetal Subjects* (University of Toronto Press, 2001).

Chapter 3 of this book and one of my articles are the only histories of amniocentesis and chorionic villus sampling that I know of: Ruth Schwartz Cowan, "Aspects of the History of Prenatal Diagnosis," *Fetal Diagnosis and Therapy* 8, supp. 1 (February 1993).

Much has been written about the history of abortion law. The most comprehensive introduction, pre–Roe v. Wade, is Daniel J. Callahan, *Abortion: Law, Choice and Morality* (Macmillan, 1970). On the movement to reform abortion law, I suggest Kristen Luker, *Abortion and the Politics of Motherhood* (University of California Press, 1984).

On Screening for PKU, Tay-Sachs Disease, Sickle-Cell Disease, and Thalassemia

Keith Wailoo has established his authority on the history of sickle-cell disease in *Dying in the City of the Blues: Sickle Cell Anemia and the Politics of Race and Health* (University of North Carolina Press, 2001), and *Drawing Blood: Technology and Disease Identity in Twentieth-Century America* (Johns Hopkins University Press, 1997), ch. 5. Wailoo and Stephen Pemberton have written tellingly about single-gene recessive diseases in *The Troubled Dream of Genetic Medicine: Ethnicity and Innovation in Tay-Sachs, Cystic Fibrosis, and Sickle Cell Disease* (Johns Hopkins University Press, 2006).

Diane Paul and Paul Edelson have done the best work on the history of newborn screening for PKU: Diane Paul, "Contesting Consent: The Challenge to Compulsory Neonatal Screening for PKU," *Perspectives in Biology and Medicine* (1999); Diane B. Paul, "The History of Newborn Phenylketonuria Screening in the U.S.," in Neil A. Holtzman and Michael S. Watson, eds., *Promoting Safe and Effective Genetic Testing in the United States: Final Report of the Task Force on Genetic Testing* (Johns Hopkins University Press, 1998); Diane Paul and Paul Edelson, "The Struggle over Metabolic Screening," in Soraya de Chadarevian and Harmke Kamminga, eds., *Molecularizing Biology and Medicine* (Harwood, 1998). Also interesting is Samuel Bessman and Judith P. Swazey, "Phenylketonuria: A Study of Biomedical Legislation," in E. Mendelsohn, Judith P. Swazey, and Irene Tavis, eds., *Human Aspects of Biomedical Innovation* (Harvard University Press, 1971).

On Tay-Sachs disease the best historical accounts have been written by the participants themselves, in Michael Kaback and Robert Desnick, eds., *Tay-Sachs Disease* (Academic Press, 2001). On thalassemia and sickle-cell disease, a relatively easy place to start is with Maxwell W. Wintrobe, ed., *Blood, Pure and Eloquent: A Story of Discovery, of People, and of Ideas* (McGraw Hill, 1980), chapters 11 and 12. For more detail on thalassemia, refer to D. J. Weatherall and J. B. Clegg, *The Thalassemia Syndromes*, 4th ed. (Blackwell Science, 2001), part 1.

Proponents and Opponents of Genetic Screening

Richard Lewontin and Jon Beckwith are the geneticists who have most passionately gone to bat against biological determinism: Lewontin, *It Ain't Necessarily So: The Dream of the Human Genome and Other Illusions* (New York Review of Books Press, 2000); Beckwith, *Making Genes, Making Waves: A Social Activist in Science* (Harvard University Press, 2002). Lewontin also coauthored *Not in Our Genes: Biology, Ideology, and Human Nature* (Pantheon, 1984) with Steven Rose and Leon Kamin. The medical geneticist who has been most articulate in defending his field is Barton Childs in *Genetic Medicine: A Logic of Disease* (Johns Hopkins University Press, 1999).

Ruth Hubbard is a biologist, left intellectual, and reproductive feminist. Her ideas are best sampled in *The Politics of Women's Biology* (Rutgers University Press, 1990) and in Ruth Hubbard and Elijah Wald, *Exploding the Gene Myth* (Beacon, 1993). There are many collections of writings by reproductive feminists, e.g. Gwynne Basen et al., eds., *Misconceptions: The Social Construction of Choice and the New Reproductive and Genetic Technologies*, 2 vols. (Voyageur, 1993), and Rita Arditti, Renate Duelli Klein, and Shelley Minden, eds., *Test-tube Women: What Future for Motherhood?* (Pandora Press, 1984). Other books by reproductive feminists include Robyn Rowland, *Living Laboratories: Women and Reproductive Technologies* (Indiana University Press, 1992); Gena Corea, *The Mother Machine: Reproductive Technologies from Artificial Insemination to Artificial Wombs* (Harper and Row, 1985); Patricia Spallone, *Beyond Conception: The New Politics of Reproduction* (Bergin and Garvey, 1989); and Joan Rothschild, *The Dream of the Perfect Child* (Indiana University Press, 2005).

For the views of disability rights activists who oppose prenatal diagnosis, see their essays in Eric Parens and Adrienne Asch, eds., *Prenatal Testing and Disability Rights* (Georgetown University Press, 2000), and in David Wasserman, Jerome Bickenbach, and Robert Wachbroit, eds., *Quality of Life and Human Difference: Genetic Testing, Health Care, and Disability* (Cambridge University Press, 2005). Feminists who favor genetic testing have rarely expressed their views in print, but some of those views are made clear in essays in these two collections, especially Eva Feder Kittay with Jeffrey Kittay, "On the Expressivity and Ethics of Selective Abortion for Disability: Conversations with My Son," in Parens and Asch.

Proponents of intelligent design are the most recent opponents of scientific materialism. To sample this literature, try Phillip E. Johnson, *Defeating Darwinism by Opening Minds* (Intervarsity Press, 1997), and Michael Behe, *Darwin's Black Box: The Biochemical Challenge to Evolution* (Simon and Schuster, 1998). The views of abortion opponents are easiest to read online at such websites as *www.michaelfund.org, www.godandscience.org, www.billmuehlenberg.com,* and *www.abortionfacts.com.*

Many bioethicists have tried to combat the opposition to genetic screening that comes from religious fundamentalists and opponents of scientific materialism. See, e.g., Philip Kitcher, *The Lives to Come: The Genetic Revolution and Human Possibilities* (Simon and Schuster, 1996), and R. M. Dworkin, *Life's Dominion: An Argument about Abortion, Euthanasia, and Individual Freedom* (Knopf, 1993). For an introduction to genetic screening for general readers that points out the pitfalls while remaining supportive, see Lori Andrews, *Future Perfect: Confronting Decisions about Genetics* (Columbia University Press, 2001); for a more enthusiastic introduction, see Doris Teichler-Zallen, *Does It Run in the Family: A Consumer's Guide to Testing for Genetic Disorders* (Rutgers University Press, 1997).

Acknowledgments

Libraries are to historians what laboratories are to scientists: we cannot do our work without them. Medical libraries create unusual challenges for historians, however, because their usual patrons face forward in time, always wanting the latest issues of journals as fast as possible, but historians face backward, always wanting the journals that have been put in storage because they have not been used for decades. I take my hat off to the staffs of the following medical libraries, who tolerated my bizarre requests patiently, sometimes even breaking into a smile at the notion that this particular patron actually wanted to read something that was not in digital format: the medical libraries at the State University of New York at Stony Brook, the University of Southern California, the College of Physicians and Surgeons of Columbia University, the University of Pennsylvania, North Shore University Hospital, the New York Academy of Medicine, the Wellcome Institute, and, most wonderful of all, the College of Physicians of Philadelphia.

Although I have not made extensive use of archival materials in the finished product, early visits to the collections of the Rockefeller Archive Center and the History of Genetics Archive at the American Philosophical Society and to the archival collections of the March of Dimes/Birth Defects Foundation were a great help in getting this project off the ground. So was the serendipitous discovery of material about Thomas Cooley in the Papers of the Detroit Children's Hospital at the Reuther Library of Wayne State University. Archival detective work done by Marcia Meldrum at the National Institute of Child Health and Human Development of the National Institutes of Health provided me with insights about the creation of the National Registry for Amniocentesis, and similar work by Julia Irwin in the libraries of Yale University helped me understand sickle-cell anemia screening programs in New Haven. On Cyprus, Michael Angastiniotis, Ayten Berkalp, and Gülsen Bozkurt were kind enough to let me copy unpublished material in their personal possession related to screening for β-thalassemia, and I am grateful to all of them for their generosity.

Over the years I have been blessed with dedicated and clever research assistants. I was ably assisted by Marcia Meldrum, Jennifer Jamilkowski, and Paula Viterbo when they were students at SUNY–Stony Brook. Years after they had gone on to careers of their own, when I read through my files in preparation for writing, I was often astonished at how thoroughly and insightfully they had summarized specialized literatures for me, making synthesis feasible. Without the assiduous detective work done by Babi Hammond I would not have been able to document either the Black Panthers' reactions to sickle-cell screening or the conflicts over contraception in African-American communities. After I finished the first draft of the manuscript—and moved to the University of Pennsylvania—Joanna Radin helped me clean up the manuscript, markedly increasing both its readability and its clarity. Matthew Hersch ably finished the cleanup, not only by tracking down recalcitrant citations (with impressive facility), but also by preparing the images for publication, an enterprise that is now done with digital tools that are completely beyond my comprehension.

Research assistants need to get paid, and scholars who are also teachers need some time off to concentrate. I am, therefore, grateful to many individuals and organizations who made it possible for me to fill both these needs: Dan Kevles, my host at CalTech when I had a Sherman Fairchild, Jr., Fellowship; the broadminded program officers at the National Science Foundation, the National Endowment for the Humanities, and ELSI (Ethical, Legal and Social Implications of the Human Genome Initiative), who arranged for collaborative funding for this project; Daniel Hadjitoffi and his staff at the Cyprus Fulbright Commission, Marina Kleanthus and her colleagues at the Cyprus Institute of Neurology and Genetics, and the professional staff members of the two thalassemia clinics in Nicosia/Lefkoşa, who provided the gracious hospitality that both eased my path and made my time productive when I was a Senior Fulbright Scholar on Cyprus in the spring of 1999. And I owe many thanks to Janice and the late Julian Bers, who endowed the research fund at Penn that supports my work.

Over the years I have interviewed dozens of physicians, parents, public health officers, patients, researchers, and technicians about the many facets of medical genetics. Although I quote from those interviews only sparingly in the text, I want to thank all these busy people for sharing their experiences and their sometimes painful memories with me. I found every interview enlightening in some way—even if, in the end, my interpretations of the meaning of events differ from those of the people who so graciously allowed me to impose on their time. Several medical experts read various chapters, and I thank them for their efforts, which have kept me from making more than one error of fact: Michael Angastiniotis, Alan R. Cohen, and Reed Pyeritz.

Finally, I want to apologize to the members of my immediate family—four people when the work on this book started, ten people (I am thrilled to be able to say) by the time it ended. I am grateful beyond measure for their patience, their encouragement, and their love.

Index

Abortion: Cypriot Orthodox Church and, 211–213; in Denmark, 93–95; ethical issues and, 240–245; eugenic, 94, 95, 108, 113; genetic testing and, 3, 4, 7, 230–231, 240–245; objections to, 7, 110–112, 144–145, 230–231; prenatal diagnosis and, 72–73, 74, 93–95, 97, 105, 110–112, 216, 239–242; reform, 107–116; Tay-Sachs disease and, 139, 144–145; terms used for, 112–113; β-thalassemia and, 206–207, 211–213, 216, 219; therapeutic, 93, 113
β-*N*-acetylhexosaminidase (Hex-A), 138–139, 141–142
Adult screening. *See* Sickle-cell anemia; Tay-Sachs disease; β-thalassemia
African Americans: birth control and, 172–173; discrimination and, 153–154, 165–166, 168–170, fitness and sickle-cell anemia, 168–169; genetic testing and, 150, 162–164; politics and sickle-cell anemia, 171–179; prevalence of sickle-cell anemia, 153
Ali, Muhammad, 163
Alkaptonuria (black urine) studies, 42–43, 47
Alleles, 52, 53
Alphafetoprotein (AFP), 102, 104
Alzheimer's disease, 3–4
Amaurotic family idiocy, 135. *See also* Tay-Sachs disease
American Eugenics Society, 20
Amniocentesis, 1, 72, 77; abortion rates and, 112; accuracy of, 103; advantages of, 97; vs. chorionic villus sampling, 107; feminism and, 240; sex-linked abnormalities and, 91, 92, 93–95; Tay-Sachs disease and, 139, 141; ultrasound and, 99
Amniotic puncture, 75
Amniotic sac (amnion), 74–75
Amniotic tap, 74–78, 77, 97, 114, 139
Angastiniotis, Michael, 200, 208, 213, 216
"Antenatal Determination of Foetal Sex in Prevention of Hereditary Diseases" (Fuchs and Riis), 92–93
Archiv für Rassen-und Gesellschaftsbiologie, 30
Armstrong, Louis, 163
Artificial insemination, 27
Asch, Adrienne, 231

Ashkenazi Jews, 117, 135, 136, 141, 144–145, 148

Bacteriology, 63
Barker, Lewellys F., 66
Barr, Murray, 87–91
Barr body (sex chromatin), 89, 90, 114
Bateson, William: argument with Karl Pearson, 56–63; background, 41, 43–45; correspondence with Archibald Garrod, 41–47; on discovering Mendel's work, 45–47; legacy of, 69; medical genetics and, 48–49; on term "genetics," 52
Bateson, William Henry, 43
Bearn, Alexander, 49
Berkalp, Ayten, 209
Bernard, Claude, 41–42
Bevis, D. C. A., 76–78, *77*, 111
Beyond Conception (Spallone), 229
Bickel, Horst, 121–122
Biochemistry, 63–64
Biological determinism, 7, 227–229, 231
Biometrika, 56
Biostatistics, 59, 60, 61, 62
Birth control: African Americans and, 172–175, 178; eugenics and, 17, 21–22, 23, 30; β-thalassemia and, 206. *See also* Sterilization
Birth control pill, and informed consent, 130
Black Panthers, 175–179, *177*
Black urine (alkaptonuria), 42–43, 47
Blood transfusions, 193–197, *194*, *196*, 199
Blood types, 62, 63, 65, 76, 77
Bozkurt, Gülsen, 216
Brave New World (Huxley), 228
Breeding studies, 45–47, *46*, 49–50, 61–62
Breg, W. Roy, 101
British eugenics movement, 15–18
Brock, D. J. H., 102
Buck, Carrie, 18–20, *19*
Buck, Pearl S., 126
Buck v. Bell (1927), 18–20, *19*

Caminopetros, J., 192
Cancer, chromosomal studies of, 79–84
Carrier screening. *See* Sickle-cell anemia; Tay-Sachs disease; β-thalassemia
Casti connubii (1930), 17
Catholicism, 17, 111
Chelation therapy, 195–196, 197, 200, 202, 209
Chemical theory of disease, 64
Chisholm, Shirley, 173
Chorion, 105
Chorionic villus sampling (CVS), 72, 105–106
Christian, Joseph, 170
Chromosomes: crossing over behavior of, 53; cytogenetics and, 78–84; discovery of, 50–51; in Down patients, 85–87, *86;* karyotyping of, 73–74, 79–82, *83;* mutations, 53, 67, 85, *86*, 101, 189; naming convention for, 82–84, *83;* prenatal diagnosis and, 99–101; sex determination and, 89
Chrysostomos, Archbishop, 211, 212
Classical genetics. *See* Mendelian genetics
Community organization: phenylketonuria and, 124–129; sickle-cell anemia and, 178–179; Tay-Sachs disease and, 136–137; β-thalasse-

mia and, 199–201, 208–209, 217–219
Contraception. *See* Birth control
Convention for the Protection of Human Rights and Dignity of the Human Being with Regard to the Application of Biology and Medicine, 223–226, 242
Cooley, Thomas B., 189–192
Cooley's anemia. *See* β-thalassemia
Corea, Gena, 230
Cosby, Bill, 163
Costs: of amniocentesis, 104; of caring for ill and disabled persons, 6, 110, 115; Medicaid and, 159; phenylketonuria testing and, 127; sickle-cell anemia testing and, 158–159, 162–163, 178; Tay-Sachs disease screening and, 142, 144; of β-thalassemia therapy, 188–189, 197, 200, 201–202, 214
Crick, Francis, 79
Cypriot Orthodox Church, 182, 201, 211–212, 219, 241
Cyprus: division of, 186–188; economy of, 188–189; history of, 182–186; infrastructure of, 183–184; mandatory screening programs, 208–212; politics and conflicts of, 184–188; politics and ethics of genetic screening, 212–222; population of, 182–183. *See also* β-thalassemia
Cyprus Antianaemic Society, 200–201, 218, 238
Cytogeneticists, 78–84
Cytological analysis, 50–51, 88

Dangerous Diagnostics (Nelkin and Tancredi), 228
Danon, Mathilde (Krim), 90
Darwin, Charles, 13, 44
Darwinism, 12, 14, 28, 44, 61, 232–234
Davenport, Charles B., 20, 65, 67, 222
Deferoxamine mesylate (Desferal), 195–196, 197, 200, 202, 209
Dermatoglyphics, 85
Desnick, Robert, 146
Deutscher Bund für Volksaufartung und Erbkunde, 31
Diet, and phenylketonuria, 121, 122–123, 124, 132
Dill-Russell, P. W., 184–185
Disability rights activists, 7, 231–232, 238–239
"Disability Rights and Selective Abortion" (Saxton), 231
Disabled persons: eugenics and, 32–34, 37–38; genetic testing and, 7, 231–232, 234, 237–239; phenylketonuria and, 118–133; Tay-Sachs disease and, 133–149
Discrimination: African Americans and, 153–154, 165–166, 168–170, 171–180; disabled persons and, 231, 243; genetic testing as, 7; phenylketonuria and, 133; β-thalassemia and, 213, 217, 220, 238–239
DNA. *See* Chromosomes
Dominant traits, 45–47
Donald, Ian, 98–99, 111
Dor Yeshorim, 146–147
Down syndrome, 84–87, *86,* 101, 104–105, 111, 112
Drosophila melanogaster (fruit fly) studies, 52–55, *54,* 69, 79
Du Bois, W. E. B., 172
Dunkell, Ruth, 139, 148
Duster, Troy, 234

Ebony, 173
Education: eugenics and, 20; Nazi eugenics and, 35–36; phenylketonuria and, *128;* sickle-cell anemia and, 167–168; β-thalassemia and, 203–204, *204, 205,* 209, 210, 213, 217
Ekstein, Josef, 145–146, 148
Electrophoresis, 157–159, *157,* 160
Ellis, "Doc," 163
Emmel, Victor E., 152–153
"Emmel test," 152, 158, 160, 168
Erythroblastic anemia. *See* β-thalassemia
Ethical concerns: decision-making and, 3–5; genetic testing and, 162–163, 223–245; informed consent and, 2, 117, 130–132; of mandatory genetic screening, 212–222; parenting and, 5, 217–219; prenatal diagnosis and, 72–73, 110–116; reproductive decisions and, 5–8. *See also* Suffering
Eugenic abortion, 94, 95, 108, 113. *See also* Abortion
Eugenicists, 15, 237; Davenport, Charles B., 20, 65, 67, 222; Galton, Francis, 12–16, 20, 26, 55–56, 67, 222; geneticists as, 55; German, 29; Gotto, Sybil, 15–16; Johnson, Albert, 21; Laughlin, Harry H., 21; Muller, Hermann J., 25–26, 67–68, 222; Pearson, Karl, 56–63, 64–65, 69
Eugenics, 9, 12–40; African Americans and, 153–155; in Britain, 15–18; in Catholic countries, 22–25; classical genetics and, 47; diversity of ideas, 39–40, 67–68; education and, 20, 35–36; effects of, 64, 66–67, 181; foundations for ideas on, 12–15; genealogical fallacy and, 67–70; human genetics and, 55–63; left-wing support for, 25–26; vs. medical genetics, 48, 49, 63–67, 234–245; Moynihan Report and, 175; and the Nazis, 28–37; opposition to, 7–8, 17, 18, 34, 40; origin of term, 13–14; physicians' attitudes and, 63–67; political origins of, 14–15; prenatal diagnosis and, 113–115; in Scandinavia, 37–40; in Soviet Union, 26–28; in U.S., 18–22, *19. See also* Human genetics
Eugenics Education Society, 15–16
Eugenics Record Office, 20, 21, 40, 65
Eugénique, 23
Evans, Dale, 126
Evolution. *See* Darwinism

Factors (allelomorphs), 51–52
Family pedigrees, 19, 55–60, 62, 192
Family planning. *See* Birth control; Marriage; Prenatal diagnosis
Fawdry, Allan, *190,* 198–199, 215–216
Feeblemindedness. *See* Eugenics; Intelligence
Feminism: abortion reform and, 108, 243–245; birth control and, 173–175; genetic testing and, 7–8, 243–245; reproductive feminists, 228–231, 237–240
Fetal sex. *See* Sex determination (fetal)
Fetoscopy, 106–107, 206–207, 208, 210, 219–220
Filipchenko, Iurii Aleksandrovich, 26
Finkbine, Sherri, 109

FINRRAGE (Feminist International Network for Resistance to Reproductive and Genetic Engineering), 228–229
Fischer, Emil, 151
Fischer, Eugen, 30
Fischer, R. A., 61, 69
Følling, Asbjørn, 118–119, 120, 123
Foundation for Research and Education in Sickle Cell Disease, 161–162
Frazier, Joe, 163
French Eugenics Society, 23
Fruit fly. *See Drosophila melanogaster* (fruit fly) studies
Fuchs, Fritz, 90, 92–95, 98, 99, 102, 114–115, 238, 240–241

Galton, Francis: background and research of, 12–15; Eugenics Education Society, 15–16; *Hereditary Genius,* 26; *Inquiries into Human Faculty,* 13–14; legacy of, 20, 67, 222; research methods of, 55–56
Galton Society, 20
Ganglioside, 136, 137
Garrod, Alfred Baring, 41
Garrod, Archibald: background and research of, 41–43, 46–48; on biostatistics, 62–63; correspondence with William Bateson, 41–47; *Inborn Errors of Metabolism,* 47; "The incidence of alkaptonuria: A study in chemical individuality," 47; legacy of, 66, 69; medical genetics and, 48–49
Garvey, Marcus, 172
Gautier, Marthe, 85
Geneticists: Bateson, William, 41, 43–49, 52, 56–63, 69; cytogeneticists, 78–84; Johanssen, Wilhelm, 52; Mendel, Gregor, 45–47, *46,* 49–50, 51, 58; Morgan, Thomas Hunt, 52–53, *54,* 58, 79; Sutton, Walter B., 51–52
Genetics: attitudes toward, 63–67; classical, 49–55, 235; early history of, 41–48, *46;* human, 55–63; medical, 41, 48–49, 155–157, 234–236
Genetic testing: African Americans and, 150; early programs in, 10–11; feminism and, 243–245; vs. genetic screening, 10; laws and support for, 6; motives of, 9, 95–96, 234–236; objections to, 7–8, 227–234; pervasiveness of, 2–4. *See also* Adult screening; Carrier screening; Newborn screening; Premarital screening; *specific genetic diseases*
Genotype, 52, 53, 62
German eugenics, 28–31. *See also* Nazi eugenics
German measles. *See* Rubella
"Germinal choice," 26
Germ theory of disease, 63, 65
Gillespie, Dizzy, 163
Goldberg, Barry, *100*
Gotto, Sybil, 15–16
Greek Cyprus. *See* Cyprus
Gregory, Dick, 173
Grotjahn, Alfred, 29
Grundriss der menschlichen Erblichkeitslehre und Rassenhygiene (Lenz, Baur, and Fischer), 30–31
Gullattee, Alyce C., 175
Guthrie, Robert, 123–124, 126–127, 129, 148
"Guthrie test." *See* Phenylketonuria (PKU)

Hadjiminas, Minas, 199–200, 208, 211, 212, 213, 216
Haldane, J. B. S., 25–26, 66
Hardy-Weinberg Law, 61, 69
Harriman, E. H., 20
Heel-prick test, 2–3. *See also* Phenylketonuria (PKU)
Hemoglobin, normal vs. sickled, 159–160
Hemophilia, 91–93, 94
Hereditary Genius (Galton), 26
Heredity: Catholic eugenicists on, 22; Galton's studies of, 13–14; sex-linked recessive traits, 58; traits and variations, 44, 45–47, 55–57. *See also* Chromosomes; *specific genetic diseases*
Heredity in Relation to Eugenics (Davenport), 26
Herrick, James B., 150–152, *151*
Hex-A (β-*N*-acetylhexosaminidase), 138–139, 141–142
History of eugenics: in Britain, 15–18; in Catholic countries, 22–25; classical genetics and, 49–55; human genetics and, 55–63; left-wing, 25–26; medical genetics and, 41, 48–49; of Nazi eugenics, 28–37; origin of, 12–15; physicians and, 63–67; in Scandinavia, 37–40; in Soviet Union, 26–28; in U.S., 18–22
Hitler, Adolf, 28, 30, 31, 32, 144
Holocaust, 144
Hopkins, F. Gowland, 42
Hsu, T. C., 79–81
Human genetics: and eugenics, 55–63; and physicians, 63–67
Human Heredity (Lenz, Baur, and Fischer), 31
Huxley, Aldous, 228
Hydrops, 76, 77

Immigration Restriction Act (1924), 20–21
Immunizations, 65
Inborn errors of metabolism, 120, 129, 132
Inborn Errors of Metabolism (Garrod), 47
"The incidence of alkaptonuria: A study in chemical individuality" (Garrod), 47
Informed consent, 2, 117, 130–132, 171, 223–224
Inheritance. *See* Heredity
Inquiries into Human Faculty (Galton), 13–14
Insurance, 103–104, 170, 239
Intelligence, 13–14, 15–16, 18
Interracial marriages, 21–22, 24–25, 154
An Introduction to the Study of Experimental Medicine (Bernard), 41–42, 44
In vitro fertilization, 225, 228

Jews, and Tay-Sachs disease, 117, 133–149. *See also* Ashkenazi Jews
Johanssen, Wilhelm, 52
Johnson, Albert, 21
Journal of Heredity, 81
Journal of Reproductive and Genetic Engineering, 229
Journal of the American Medical Association (JAMA), 162–163
Journal of the National Medical Association, 175

Kaback, Michael, 140–141, 143, 148, 149
Kaiser, I. H., 90
Karyotyping, 73–74, 78, *86,* 93, 101, 114
Kennedy, John F., 126

Kennedy, Rosemary, 126
Kernicterus, 76
King, Colby, 164
King, Martin Luther, Jr., 175
Klenk, Ernst, 136
Koltsov, Nikolai Konstanovich, 26
Krim, Mathilde (Danon), 90

Lamarck, Jean-Baptiste, 23
Lancet, 89, 90, 101, 113
Land, Theobald, 32
Laughlin, Harry H., 21
Law for the Prevention of Genetically Diseased Offspring (1933), 32–34
Law for the Protection of the Genetic Health of the German People (1935), 34–35
Law for the Protection of the German Blood and German Honor (1935), 34–35
Lee, Pearl, 189–191
Left-wing intellectuals, 7, 227–228
Legislation and legal issues: abortion and, 6, 93–95, 105, 107–116; *Buck v. Bell* (1927), 18–20, *19;* Convention for the Protection of Human Rights and Dignity of the Human Being with Regard to the Application of Biology and Medicine, 223–226; immigration policy and, 20–21; informed consent, 2, 117, 130–132, 171, 223–224; Law for the Prevention of Genetically Diseased Offspring (Germany, 1933), 32–34; Mental Deficiency Act (Britain, 1913), 17; National Sickle-Cell Anemia Control Act (U.S., 1972), 164–166, 167; newborn screening and, 129; Nuremberg Laws (Germany, 1935), 34–35; phenylketonuria testing and, 127–129; premarital health exams (France, 1942), 23–24; Social Security Act and family planning (1967), 172; sterilization in Scandinavia, 38–39; sterilization in U.S., 18; Turkish Family Law, 209–210
Lejeune, Jérôme, 84–87, 89, 111–112
Lenz, Fritz, 30, 31
Levan, Albert, 81–82
Lévi-Strauss, Claude, 78
Liberator, 172
Life, 99
Lilley, William, 111
Lippman, Abby, 228–229
Los Angeles Sickle Cell Anemia Foundation, 178–179
Luker, Kristin, 111
Lysenko, Trofim, 27–28

MacIntyre, M. Neil, 241–242
Makarios, Archbishop, 185–186
Makowski, E. L., 90
March of Dimes/Birth Defects Foundation, 84, 90, 111, 125, 136
Marcus Welby, M.D. (television show), 164
Marriage: eugenics and, 21–22, 23, 34–35; interracial, 21–22, 24–25, 154; sickle-cell anemia screening and, 162, 165; Tay-Sachs screening and, 139, 142, 144, *145,* 146; β-thalassemia and, 181–182, 183, 203, *204, 205,* 206, 209–210, 220–221
The Mechanism of Mendelian Inheritance (Morgan), 53, 54, 79
Medicaid, 159, 165
Medical care: on Cyprus, 188–189; eugenics and, 28–30, 32–34, 63–67; Medicaid and, 159, 165; screening and, 218. *See also* Costs

Medical genetics, 41, 48–49, 66, 155–157, 234–245, 236. *See also* Genetics; Prenatal diagnosis
Mediterranean ancestry, 191
Meiosis, 50–51
Mendel, Gregor, 45–47, *46,* 49–50, 51, 58
Mendelian genetics, 57–58, 235. *See also specific genetic diseases*
Mental Deficiency Act (Britain, 1913), 17
Mitosis, 50
Modell, Bernadette, 209
Mongolism. *See* Down syndrome
Morgan, Thomas Hunt, 52–53, *54,* 58, 79
Mother's Aid Institution (Denmark), 94
Moynihan, Daniel Patrick, 173–176
Moynihan Report, 173–176
Ms. Magazine, 240
Muehlenberg, Bill, 233
Muller, Hermann J., 25–26, 67–68, 222
Murray, Robert, 169
Mutation, 53, 67, 85, *86,* 101, 189

National Association of Parents and Friends of Mentally Retarded Children (NARC), 124–129, 148, 238
National Office of Social Hygiene (France), 23–24
National Sickle-Cell Anemia Control Act (U.S., 1972), 164–166, 167
National Socialists. *See* Nazi eugenics
National Tay-Sachs Disease Association (NTSDA), 136, 139, *145,* 148, 238
Nature, 82, 89
Nature vs. nurture, 13–14, 60
Nazi eugenics, 28–37; education and, 35–36; effects of, 181, 225; marriage, 34–35; medical care, 28–34; murder, 36–37; research, 30–31; Tay-Sachs screening and, 147–148
Neel, James, 69–70, 71, 155–157, 158, 192–193, 236
The Negro Family: The Case for National Action (Moynihan), 173–176
Nelkin, Dorothy, 228
Neo-Darwinian synthesis, 61, 62
Neural tube defects, 102, 107, 111
Newborn screening: critics of, 129–133; history of, 127–129. *See also* Phenylketonuria (PKU); Screening programs
New England Journal of Medicine, 168
Nixon, Richard, 164–165, 176–177
Nuclear satellite, 88–89
Nuremberg Laws (Germany, 1935), 34–35

O'Brien, John S., 139, 140
Okada, Shintaro, 139
Oligohydramnios, 75
On the Origin of Species (Darwin), 13, 44
"Oreos" (derogatory term), 176
Organ donation, 225
Osmotic fragility test, 198

Painter, T. S., 79–80
Parental consent. *See* Informed consent
Parenthood, ethics of, 217–219, 245. *See also* Reproductive decisions
Paternity (disputed), 65
Pauling, Linus, 157–158, *157*
Pearson, Howard, 166
Pearson, Karl, *56–63,* 64–65, 69
Penrose, Lionel, 40, 119–121, 124

Petricelli, Leonard, 164
Petricelli, Robert, 164
Phenotype, 52, 53, 62
Phenylalanine, 119, 121
Phenylketonuria (PKU), 2–3, 118–133, *128;* diagnostic test for, 123–124; politics and, 124–129; pregnancy and, 132; screening objections, 129–132; screening today, 148–149
Phenylpyruvic acid, 119, 121. *See also* Phenylketonuria (PKU)
Physicians: abortion reform and, 108, 111; on Cyprus, 187–189; human genetics and, 63–67; phenylketonuria and, 118–123; sickle-cell anemia and, 167–171; Tay-Sachs disease and, 133–135, 140–141; β-thalassemia and, 203–204, 207–208, 215–216
Ploetz, Alfred, 29, 30, 31
Poitier, Sidney, 163
Politics: of Black Panthers, 175–179, *177;* of British eugenics, 15–18; Catholicism and, 22–25; of left-wing eugenics, 25–26; of Nazi eugenics, 28–37; of parenthood, 5–8; of PKU screening, 124–129; of Scandinavian eugenics, 37–40; sickle-cell anemia and, 163–164, 171–179, *177;* of Soviet eugenics, 26–28; Tay-Sachs disease and, 136–138; of β-thalassemia screening, 212–222; of U.S. eugenics, 18–22
Polyhydramnios, 75, 78
Poverty (pauperism), 15–18
Powell, James A., 168
Pregnancy: medicalization of, 7–8, 72, 113, 220, 229; phenylketonuria and, 132. *See also* Abortion; Prenatal diagnosis
Prem, K. A., 90
Premarital screening: for sickle-cell anemia, 165, 166, 171; for Tay-Sachs disease, 144–147, *145;* for β-thalassemia, 203–206, *205,* 208–212
Prenatal diagnosis, 10–11, 71–116; abortion reform and, 107–116; amniotic tap for, 74–78; demand for, 96, 102; Down syndrome and, 87; ethical issues and, 72–73, 234–245; vs. eugenics, 113–115; goals of, 236–237; innovation in, 95–107; motivations in, 95–96; objections to, 110–112; research in, 84–87; Rh disease and, 75–78; sex chromatin and, 87–93; sickle-cell anemia and, 165, 180; Tay-Sachs disease and, 139; as technological system, 73–74, 95–107; β-thalassemia and, 203, 206–207, 210. *See also* Alphafetoprotein (AFP); Amniocentesis; Chorionic villus sampling (CVS); Fetoscopy; Ultrasound
Prevention: abortion role in, 93–95, 112; cost-benefit of, 214; of Tay-Sachs disease, 138–139, 146–147; of β-thalassemia, 202–212, 214. *See also* Genetic testing; Screening programs
Probability, 57–60
Pro-lifers, 7, 110–112, 232–234, 240
Public policy. *See* Legislation and legal issues; Politics

Racism. *See* Discrimination, Eugenics; Interracial marriages
Rapp, Rayna, 9
Rassenhygiene, 29
Recessive traits, 45–47
Reduction division, 50–51

Religious fundamentalism, 232–234
Reproductive decisions: sickle-cell anemia and, 171–172; social aspects of, 5–8; Tay-Sachs disease and, 146–147; β-thalassemia and, 203–206. *See also* Abortion; Birth control; Marriage; Prenatal diagnosis; Prevention
Reproductive feminism, 7–8, 228–231, 237–240
Republic of Cyprus. *See* Cyprus
Research: on alphafetoprotein (AFP), 102; on Alzheimer's disease, 3–4; on amniocentesis, 102–103; on black urine (alkaptonuria), 42–43, 47; on chorionic villus sampling, 105–107; by clinicians, 84–87; on cytogenetics, 50–51, 78–84; on Down syndrome, 84–87; and medical genetics, 71–72; on Mendelian genetics, 54–55; on obstetric ultrasound, 104–105; on phenylketonuria, 118–124; on Rh disease, 75–78; on sex chromatin, 87–93; on sickle-cell anemia, 156–157, *157,* 164, 169; on Tay-Sachs disease, 133–138; on β-thalassemia, 189–192, *190;* Tuskegee syphilis study, 171; validity of methods used for, 57–63
RhD factor, 76–78, 77
Rh disease, 75–76
Riis, Povl, 90, 92–95, 98, 99, 102, 114–115, 240–241
Risk. *See* Safety; *specific genetic diseases*
Rogers, Roy, 126
Rothman, Barbara Katz, 230
Rubella, 109–110
Russian Eugenics Society, 26
Russkii Evgenischeskii Zhurnal, 26
Sachs, Bernard, 133–134
Sachs, Leo, 90
Safety: amniocentesis and, 99, 102–103; of blood transfusions, 194–195; chorionic villus sampling and, 105–107; fetoscopy and, 207, 220
Sanger, Margaret, 172
Saxton, Marsha, 231
Scandinavian eugenics, 37–40
Schallmeyer, Wilhelm, 29
Scientific materialism, 232–233
Scott, Robert B., 162–163, 164
Screening programs: failure of, 179–180; mandatory, 127–133, 208–222; objections to, 118, 129–132; 144–147; for phenylketonuria, 126–129; for sickle-cell anemia, 150, 159, 160–167, 178–180; stigmatization and, 169–171; success of, 146–147, 148–149, 207–208, 211, 212; for Tay-Sachs disease, 140–143, *145,* 146–147; for β-thalassemia, 202–212, *204, 205;* tuberculin, 117–118. *See also* Carrier screening; Genetic testing; Newborn screening; Premarital screening
Seaman, Barbara, 240
Serebrovskii, Aleksandr, 27
Serr, David, 90
Sex chromatin (Barr bodies), 89, 90, 114
Sex determination (fetal), 87–90, 105
Sex-linked abnormalities, 91
Shettles, Landrum, 91
Sickle-cell anemia, 150–180, *151;* conflict and, 166–171; diagnostic tests for, 106, 152–153, 154, 157–161, *157;* discovery of, 150–152; eugenic beliefs and, 153–155; latent vs. patent, 153, 167, 170–171; politics of, 163–164, 171–179,

177; research on, 155–158, *157;* screening and, 11, 161–167; symptoms and signs of, 155
Sickle Cell Disease Association of America, 150
Sickledex, 159–161, 163, 166, 176–178
Slome, David, 135–136
Snyder, Laurence, 66
Social Security Act, 172
Spallone, Patricia, 229, 230–231
Spina bifida, 102, 107, 111
Stalin, Joseph, 27–28
Stamatoyannopoulos, George, 202–203, 208, 214
Statistics, and probability, 57–60
Steele, Mark W., 101
Sterilization: British eugenics movement and, 17; in Germany, 32–34; pedigree chart for, *19;* Lionel Penrose on, 119–120; in Scandinavia, 37–38; in Soviet Union, 27; in U.S., 18–20
Suffering, 239; medical genetics and, 236, 243–245; phenylketonuria and, 132–133; prenatal diagnosis and, 115–116; sickle-cell anemia and, 155, 158, 164; Tay-Sachs disease and, 139–141; β-thalassemia and, 197, 198, 211, 216, 217, 218–220
"Survival programs," 176–177
Sutton, Walter B., 51–52
Sweet pea genetics, 45–47, *46*

Tancredi, Laurence, 228
Tay, Warren, 133
Tay-Sachs disease, 133–149; cause of, 137–138; naming of, 133, 136; prevention of, 138–139; research about, 136–138; success of, 149; symptoms and signs of, 133–135; testing and screening for, 118, 140–143, *145,* 146–147. *See also* Dor Yeshorim
Technological systems, 73–74
Technologies: of amniocentesis, 71–74, 104; amniotic tap, 74–78; blood transfusions, 193–197, *194, 196,* 199; chorionic villus sampling and, 105–107; of cytological analysis, 50–51; electrophoresis, 157–159, *157,* 160; fetoscopy, 106–107, 206–207, 208, 210, 219–220; ophthalmoscope, 133; osmotic fragility test, 198; prenatal diagnosis and, 71–74, 95–107; Sickledex testing, 159–161, 163, 166, 176–178; ultrasound, 72, 98–99, *100,* 104; wet prep (Emmel test), 152
Thalassaemia Association, 209, 210
β-thalassemia, 11, 106, 155–156, 181–222; diagnostic test for, 198; identification card for, *204, 205;* incidence of (Cyprus), 205–206, 212; mutation for, 189; naming of, 192; prevalence of (Cyprus), 198, 202; prevention program for, 202–208; research on, 189–192, *190;* symptoms and signs of, *190,* 191, 194; treatment for, 193–197, *194, 196. See also* Cyprus
Thalassic anemia. *See* β-thalassemia
Thalidomide, 108–109
Therapeutic abortion, 93, 113. *See also* Abortion
Tijo, Joe-Hin, 81–82
Transfusions. *See* Blood transfusions
Treatment: for phenylketonuria, 121–122, 129–132; for Tay-Sachs disease, 138; for β-thalassemia, 193–197, *194, 196,* 199, 200, 202, 209
Trisomy 21. *See* Down syndrome

Turkish Family Law, 209–210
Turkish Republic of Northern Cyprus. *See* Cyprus
Turner, Frank Douglas, 121
Turpin, Raymond, 84, 85
Tuskegee syphilis study, 171
Twins, 3, 60

Ultrasound, 72, 98–99, *100,* 104
United States, eugenics movement in, 18–22
Universal Negro Improvement Association, 172
U.S. Department of Defense, 168–169
U.S. Department of Health, Education, and Welfare (HEW), 164–165

Valentine, William, 192–193
Variation. *See* Chromosomes; Mutation

Watson, James, 79
Webb, Z. Ozella, 159
Weismann, August, 53
"Wet diaper" test, 122–123
"Wet prep" test, 152, 158, 160, 168
Whipple, George, 192